SUPPLYING THE MINING WORLD:

The Mining Equipment Manufacturers of San Francisco, 1850–1900

SUPPLYING THE MINING WORLD:

The Mining Equipment Manufacturers of San Francisco, 1850–1900

By

Lynn R. Bailey

WESTERNLORE PRESS . . . 1996 . . . TUCSON, ARIZONA

Library of Congress Catalog Card Number 96-60529
ISBN 0-87026-096-0

PRINTED IN THE UNITED STATES OF AMERICA BY WESTERNLORE PRESS
Post Office Box 35305, Tucson, Arizona 85740

Table of Contents

Foreword

THE California Gold Rush drew tens of thousands of men to the Pacific Coast. Most who came knew nothing of mining; nevertheless, they hoped to strike it rich. Others sent by large mining companies to develop auriferous deposits, were knowledgeable of mining technology, schooled in Europe, South and Central America, and on the eastern seaboard of the United States. They brought hydraulic processes, rock crushing techniques, and amalgamation. Still another breed of men responded to the demands of commerce and transportation: mechanics, millwrights, blacksmiths, iron and brass founders, boiler makers, cordage and wire rope makers, engineers. Thus by 1850 there existed on the Pacific Coast a pool of talent which would soon be put to good use.

California's isolation, coupled with complex mineralogical problems, which in many cases had to be solved instantly, decreed that there would develop on the West Coast an industry which would make San Francisco a center for advancement of mining technology. In dozens of blacksmith shops, foundries, and crude laboratories along the town's waterfront Old World mineral extraction processes were tested and refined. Current technology was combined with good old Anglo intuition to turn mining appliances and processes into something that was uniquely American. An age-old European rock crusher was repeatedly redesigned to fit ever-changing mineralogy. Mechanical ore feeders were added. The result was the birth of the California Quartz Mill, a versatile and dependable machine that would be used worldwide to extract billions of dollars worth of gold and silver. At first these mills were water powered, then fine steam engines were perfected to drive them and their auxiliary equipment. That was just the beginning. Harnessed to hoists and pumps, these powerplants solved haulage and drainage problems at Grass Valley, the Mount Diablo coal fields, and on the Comstock. The profitable exploitation of silver-lead and copper carbonate deposits throughout Western North America is attributable to the portable water-jacket furnace, a unique San Francisco invention. The hydraulic nozzle and the Pelton water-wheel also originated on the West Coast. Little wonder that mining men from all parts of the world came to San Francisco to obtain what they wanted for their enterprises.

In the short span of a decade forty-seven foundries and machine shops had sprouted on land reclaimed from the sea. Most started as maritime repair facilities, but the demand for dependable mining equipment permanently cast these companies into the role of mining equipment suppliers. By 1875 the Union Iron Works, Pacific Foundry, the Miner's Foundry, Golden State, Fulton, Aetna, Vulcan, Risdon, Parke and Lacy, Joshua Hendy, the California Wire Works, and the Pelton Company were names known to mining men around the world. These and lesser enterprises created, tested, and marketed every variety of appliance needed to extract precious metals. Their products were shipped to all United States mineral producing districts, as well as to Canada, Mexico, Central and South America, Australia, New Zealand, Japan, China, Russia, and other countries around the Pacific Rim. San Francisco literally supplied the Mining World.

San Francisco's manufacturing monopoly lasted barely twenty years. The lifting of the wall of isolation killed the commerce. Eastern competition came when the continent was spanned by railroads. Companies such as Fraser and Chalmers, forerunner of Allis-Chalmers, grew rapidly marketing mining equipment built from plans pirated from San Francisco enterprises. While railroads made it easy to transport equipment, tariffs syphoned profits, and by the 1890s San Francisco foundries and machine shops either closed or changed direction. The Union Iron Works and Risdon became premier shipyards, fabricating naval vessels through two world wars. In the end, they too were gobbled up by Eastern capitalists.

For the most part, however, the companies that made San Francisco a common denominator in the mining industry are forgotten. Except for a few rusting relics in remote mining districts or in museums, their equipment have been scrapped long ago. Had it not been for one San Francisco newspaper, the *Mining and Scientific Press,* the history of that city's mining equipment manufacturers would have been lost.

From its first issue in May of 1860 the *Mining and Scientific Press* devoted attention to the emerging foundry trade. With development of the Comstock Lode that coverage increased until by the 1870s the paper was publishing a weekly column entitled "Foundry Notes," which detailed what each foundry was doing, the type of equipment they produced, its uniqueness, and the destination of the appliances. The articles also presented great historical insight. Growth of foundries were outlined, and biographical sketches of the principal owners appeared frequently. In many instances the articles were illustrated with line drawings of the establishments and their equipment, taken from patent applications, manufacturers' catalogs, and from mining publications. Every week for twenty years the *Mining and Scientific Press* published its Foundry Notes — a valuable and overlooked contribution to the history of American invention and mining technology.

While researching mining topics I became enthralled with the Foundry Notes. What a wonderful publication they would make if combined with the line drawings. Thus began the search. I collected the articles, rewrote them, adding additional information gleaned from other mining and historical sources. The search for quality photographs of equipment took me to other publications. Still others were found in mining memorabilia supplied by rare book dealers. In most cases I was able to duplicate the illustrations appearing in the *Press.* Through the help of the University of Arizona Library, the Arizona State Historical Society, and collector-historian John D. Gilchriese, I eventually assembled enough illustrations to round out the story of San Francisco's mining equipment manufacturers. This book is the result. I hope you will enjoy the presentation and that it may cause you to delve further into an all but forgotten phase of Pacific Coast history.

Lynn R. Bailey
Tucson, Arizona
February 6, 1996

SUPPLYING THE MINING WORLD:

The Mining Equipment Manufacturers of San Francisco, 1850–1900

Sutter's Mill, where James W. Marshall made his gold discovery of January 24, 1848, the first practical recovery of precious metal in California, which led to the "excitement" of 1849. As this engraving shows, Sutter's Mill was situated in a pleasant valley on the south fork of the American River, in El Dorado County, for a time known as Empire County. (*Mining and Scientific Press*, July 2, 1881)

CHAPTER ONE

Roots of the Foundry Trade: A Renaissance Rock Crusher and California's Gold Rush, Quartz Mining, and the Comstock Lode

GOLD was known to exist in California at least three decades before it turned up in a mill race at Coloma on the south fork of the American River. In his 1816 book, *A System of Mineralogy*, published in Edinburgh, Scotland, Robert Jameson alludes to gold in California. But his claim is unsubstantiated, for he gives no location. He may have been referring to the Pot Holes, on the west bank of the Colorado River about ten miles above present-day Yuma, where Spaniards gleaned some gold in the late 1700s.

The first verifiable report of gold, however, dates from 1825, when a party of trappers headed by Jedediah S. Smith, found gold "in quantities" near Mono Lake. In 1843 Manuel Castañares, deputy from California to the Mexican Congress, reported 2,000 ounces of gold in circulation at Los Angeles. This placer metal was obtained from Santa Feliciano Canyon, forty miles northwest of the pueblo and possibly from deposits along streams flowing out of the San Gabriel Mountains.[1] None of these reports and discoveries set off a stampede to the west coast of North America.

The January 24, 1848, discovery of gold by James W. Marshall, a wheelwright from New Jersey, was different. It was, as T. A. Rickard observed, "the most portentous event in the history of modern mining." This discovery of commercial quantities of the yellow metal set off a chain of worldwide events — migration, expansion of international trade, and scientific inquiry and technological innovation.

The discovery of a few grains of gold in the tail race of John Sutter's mill, nine days before California was ceded by Mexico to the United States by virtue of the Treaty of Guadalupe Hidalgo, could not be kept secret. Chaos reigned once the word was leaked. Employees at the mill left to dig gold; men on Sutter's hacienda decamped leaving his crops unharvested in the fields; his cattle and horses were stolen. On March 15, 1848, the San Francisco *Californian* announced the discovery. On May 29 the paper stretched the truth by reporting "the whole country, from San Francisco to Los Angeles, and from the seashore to the base of the Sierra Nevada, resounds with the sordid cry *gold! gold! gold!*"

News of the discovery spread rapidly throughout California. Soldiers deserted the presidio. San Francisco, a town of about 900 population at the time of Marshall's discovery, dropped to less than a hundred in late June. Monterey was vacated. The scene was repeated up and down the coast. Rural areas were no exception: farmers left their fields unattended, ranchers abandoned their cattle. Everyone gripped picks, shovels, and every variety of pan, and raced for mountain streams to wash gold.

Trading ships carried news of the discovery to the

Sandwich Islands, the Hawaiian Islands of today. From these Pacific crossroads news was relayed to China, Australia, around the Pacific Rim. Mexico and the Latin American countries likewise received the news from traders. It took a few months for published accounts to reach the Atlantic seaboard of the United States. By end of summer the word was out and by December 1848 gold fever had swept the world. Men were seized with the vision of enrichment; of sharing in the spoils of El Dorado, if not by personal participation, then by the indirect method of trade.

The first ships to arrive in California following the gold discovery came from South American ports; cargos from the Atlantic seaboard followed, until ships in San Francisco's harbor were counted by the hundreds. In the last nine months of 1849 more than 700 vessels docked; over 850 were logged for the year ending June 1851. Many cargos were left in the holds to decay as crews, smitten by gold fever, hurried to the diggings. The abandoned vessels rotted and sank in the harbor. Others were sold for port dues, or converted into stores and lodging houses.

Gold fever continued aboard long after the eventful year of 1849, and people continued to pour into California. Most came by sea: 36,000 in 1850, 27,000 in 1851; reaching a climax in 1852 with the arrival of 67,000. Owing to the number of routes, overland migration can only be vaguely estimated. Probably 40,000 came by the central route in 1850, and 15,000 passed through Arizona and Mexico.[2] It was the "greatest world migration in history to any one single spot for any one single purpose."[3]

Californians close to the initial discovery of gold were, for the most part, ignorant of the principals of geology and mining, and failed to recognize that placer gold had been derived by erosion from quartz veins in nearby hills. They merely worked streams by whatever means at hand. That would change with arrival of new technology. Among the flood of humanity racing across the Plains, over the Isthmus of Panama, and around Cape Horn, were men who had worked gold-bearing streams and rock in Mexico, Central America, Europe, and the southern Appalachian region of the United States. It was miners from the latter area, specifically from Georgia and the Carolinas, who brought the cradle or rocker, the long-tom, riffled sluices, much hydraulic equipment, and the stamp mill — appliances which gave the first impulse to gold mining in California.[4] Knowledge and intuition told these men to look for gold-bearing quartz veins along the lower slopes of the Sierra Nevada.

Within eighteen months of Marshall's discovery, there was an organized effort underway to locate the source of placer gold. In winter of 1850–51 a company was formed in Placerville, El Dorado County, to explore for quartz veins. The region chosen for exploration was from Mathenas Creek west on both sides of the north fork of the Cosumnes River, as far west as Shingle Springs. The search went on without success until April 1851, at which time the group discussed disbanding. They agreed to one last exploration. The men would go in pairs, taking three different but roughly parallel routes, returning to camp by the trails, and sharing the results, if any, equally. Their final effort paid off. The last pair of men to return to camp on April 12, 1851, reported finding a large outcrop of gold-bearing quartz. The discovery resulted in the development of the Union Quartz Mine.

With 100,000 men seeking gold in the foothills of the Sierra Nevada, it was natural that gold-bearing quartz would be discovered elsewhere. Such was the case at Grass Valley, then called Centerville; only that find was made accidentally. In September 1850 a German, his name is unknown, while carrying a bucket of water disturbed with his foot a piece of quartz on Gold Hill. Recognizing that it contained gold, he sold it to a shopkeeper for five dollars, who in turn unloaded the specimen for $100. News of the find spread, drawing placer miners from nearby Deer Creek. All hurried to the hill where they broke up the exposed quartz. They found nothing and the excitement died away for a few months.

On November 1, 1850, a party of miners while quarrying rock on Gold Hill for a chimney of a cabin they were building, struck a rich gold-bearing quartz vein. More gold streaked quartz was found on nearby Massachusetts Hill. Claims were immediately staked off and the men went to work. At first they broke the quartz with hammers and pryed the gold from its matrix with knives. Then they pounded the quartz in mortars with rocks attached to spring poles; Mexican arrastras were tried next. It was hard, back-backing work. By these primitive methods it took a week to pulverize two tons of quartz. But it was not long before enterprising men appeared with other means to mechanically crush the ore.

The first to sense an opportunity was Col. John C. Fremont, the great pathfinder who first appeared in California in 1842–43. He procured a small steam engine and mill of eight stamps in the East and shipped it around the Horn. Under direction of James Duff, this mill was set up on Mariposa Creek in August 1850. Although unsuccessful, it was California's first quartz mill. In 1851 Commodore Robert Stockton put

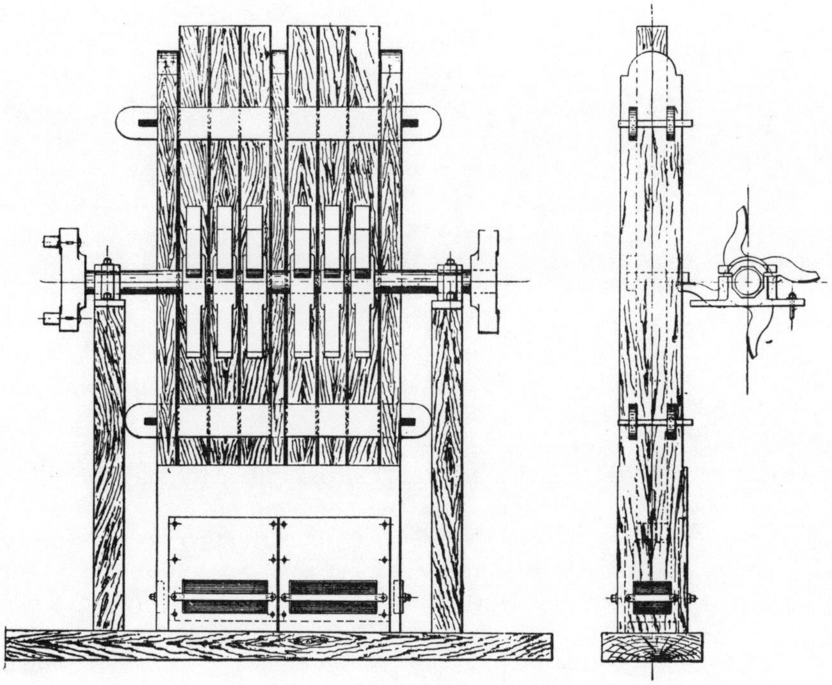

At left: The Premium Battery, a gravity stamp mill shipped from England and erected in Grass Valley in 1853 by the Agua Fria Company. The mill was brought into production under the supervision of Melville Attwood and Almarin B. Paul, two metallurgists and millmen who formulated the Washoe process which put the Comstock on a paying basis. (*Mining and Scientific Press*, May, 25, 1895)

Below: Grass Valley in 1853. The mining world is indebted to this gold producing area for the initial advances in techniques of milling and lode mining. Demand for machinery from Grass Valley and Nevada City transformed San Francisco foundries into equipment manufacturers. (*Mining and Scientific Press*, June 8, 1895)

up two quartz mills, one on Stockton Creek a mile from the town of Mariposa, and the other in Grass Valley. These two water-powered mills also failed. About the same time, two Germans built another water-powered mill on Wolf Creek, also in Grass Valley. And Messrs. Sowers, Abby & Ridgell erected a small steam-powered mill, and James Walsh brought in another quartz mill that spring. Successful, this mill was sold to Collins & Crossett, and had a long history of payment.

While all this was going on, the original Gold Hill Company (the men erecting the cabin) had pryed about $20,000 worth of gold from the quartz; enough to justify the erection of another mill of 18-stamps. A controlling interest in this mill and its claims was purchased by the Agua Fria Company, a British partnership, in August 1852. The following year the company erected a 21-stamp mill. More mills followed as miners focused on quartz mining: the New Helvetia mill of Conway & Preston in fall of 1851; Col. Rich-

ardson's mill and the Mount George mill in summer of 1851; the French Company mill, the Rocky Bar Mining Company mill, and the mill of James Winchester in 1852. And in 1853 the Union Company put up its mill. That year twenty-one stamp mills were erected in California. Five years later there were 280 quartz mills, with a total of 2,610 stamps, representing an investment in construction cost alone of more than $3,000,000. The era of quartz milling had dawned.

What was so special about these mills? In 1850–53 nothing. Their construction followed the same plan as stamp mills used in the gold-belt of the southern Appalachian region of the United States since the early 1830s. No more than a set of trip hammers; and those devices have been around for nearly 2,000 years. The Chinese were hulling rice with water-driven trip hammers as early as A.D. 290.[5] But we are concerned with rock not rice, interested in stamp mills as they apply to Western mining and the growth of San Francisco's machinery manufacturing companies.

Regardless of their Asian roots, we want to know when stamps mills were first used in Europe to crush rock, and by whom. In all likelihood the Germans were the first to use stamps to crush rock. A Saxon nobleman named Von Maltitz was milling ore with stamps about 1505; and Paul Grommestetter, a native of Schwarz, introduced the stamp mill at Schneeberg; and in 1521 he established a process of wet stamping and sifting at Joachimsthal. Grommestetter enlarged the latter plant two years later. Recent archaeological evidence, however, indicates these men were newcomers to the European milling scene.

Trip hammer appliances are associated with the forges of Schmidmulem in A.D. 1010, and were being used to mill ore all over Europe by the 13th Century. Stamp mills were crushing cassiterite, the ore of tin, in Cornwall as early as 1402.[7]

Without a doubt stamp mills were Medieval devices, but it is not until the Renaissance that a picture emerges of their components and operation. A Saxon physician, Georgius Agricola (his given name was Georg Bauer), when not practicing medicine was a frequent visitor to Freiberg and other German mineral producing locales. While studying mineralogy, he observed and sketched mining practices and equipment. His *De Re Metallica,* published in 1556, gives us the earliest description of the stamp mill. At this point it is best Agricola speaks for himself.

Ore is crushed with iron-shod stamps, in order that the metal may be separated from the stone and the hanging-wall rock. The machines which miners use for this purpose are of four kinds, and are made by the following method. A block of oak timber six feet long, two feet and a palm square, is laid on the ground. In the middle of this is fixed a mortar-box, two feet and six digits long, one foot and six digits deep; the front, which might be called a mouth, lies open; the bottom is covered with a plate of iron, a palm thick and two palms and as many digits wide, each end of which is wedged into the timber with broad wedges, and the front and back part of it are fixed to the timber with iron nails. To the sides of the mortar above the block are fixed two upright posts, whose upper ends are somewhat cut back and are mortised to the timbers of the building. Two and a half feet above the mortar are placed two cross-beams joined together, one in front and one in the back, the ends of which are mortised into the upright posts already mentioned. Through each mortise is bored a hole, into which is driven an iron clavis; one end of the clavis has two horns, and the other end is perforated in order that a wedge driven through, binds the beam more firmly; one horn of the clavis turns up and the other down. Three and a half feet above the cross-beams, two other cross-beams of the same kind are again joined in a similar manner; these cross-beams have square openings, in which the iron-shod stamps are inserted. The stamps are not far distant from each other, and fit closely in the cross-beams.

Each stamp has a tappet at the back, which requires to be daubed with grease on the lower side that it can be raised more easily. For each stamp there are on a camshaft, two cams, rounded on the outer side, which alternately raise the stamp, in order that, by its dropping into the mortar, it may with its iron head pound and crush the rock which has been thrown under it. To the camshaft is fixed a water-wheel whose buckets are turned by water-power.

Instead of doors, the mouth of the mortar has a board, which is fitted into notches cut out of the front of the block. This board can be raised, in order that when the mouth is open, workmen can remove with a shovel the fine sand, and likewise the coarse and broken rock, into which the rocks have been crushed; this board can be lowered, so that the mouth thus being closed, the fresh rock thrown in may be crushed with the iron-shod stamps. If an oak block is not available, two timbers are placed on the ground and joined together with iron clamps, each of the timbers being six feet long, a foot wide, and a foot and a half thick. Such depth as should be allowed to be the mortar, is obtained by cutting out the first beam to a width of three-quarters of a foot and to a length of two and a third and twenty-fourth of a foot. In the bottom of the part thus dug out, there should be laid a very hard rock, a foot thick and three-quarters of a foot wide; about it, if any space remains, earth or sand should be filled in and pounded. On the front, this bed rock is covered with a plank; this rock when it has been broken, should be taken away and replaced by another. A smaller mortar having room for only three stamps may also be made in the same manner.

Agricola's Renaissance gravity stamp mill. The mill's components are identified as follows:
A: Mortar
B: Upright posts
C: Cross-beams
D: Stamps
E: Their heads
F: Axle, or camshaft
G: Tooth of the stamp, or tappet
H: Teeth of axle (cams)

The stamp-stems are made of small square timbers nine feet long and a half a foot wide each way. The iron head of each is made in the following way; the lower part of the head is three palms long and the upper part the same length. The lower part is a palm square in the middle for two palms, then below this, for a length of two digits it gradually spreads until it becomes five digits square; above the middle part, for a length of two digits, it again gradually swells out until it becomes a palm and a half square. Higher up, where the head of the shoe is enclosed in the stem, it is bored through and similarly the stem itself is pierced, and through the opening of each, there passes a broad iron wedge, which prevents the head falling off the stem. To prevent the stamp head from becoming broken by the constant striking of fragments of ore or rock, there is placed around it a quadrangular iron band a digit thick, seven digits wide, and six digits deep. Those who use three stamps, as is common, make them much larger, and they are made square and three palms broad each way; then the iron shoe of each has a total length of two feet and a palm; at the lower end, it is hexagonal, and at that point it is seven digits wide and thick. The lower part of it which projects beyond the stem is one foot and two palms long; the upper part, which is enclosed in the stem, is three palms long; the lower part is a palm wide and thick; then gradually the upper part becomes narrower and thinner, so that at the top it is three digits and a half wide and two thick. It is bored through at the place where the angles have been somewhat cut away; the hole is three digits long and one wide, and is one digit's distance from the top. There are some who make that part of the head which is enclosed in the stem, barbed and grooved, in order that when the hooks have been fixed into the stem and wedges fitted to the grooves, it may remain tightly fixed, especially when it is held with two quadrangular iron bands. Some divide the cam-shaft with a compass into six sides, others into nine; it is better for it to be divided into twelve sides, in order that successively one side may contain a cam and the next be without one.[8]

Linguistic data indicate that stamp mills were introduced to the British Isles from the Continent, although Cornishmen like to think it was the other way round. According to geologist Herbert Hoover, translator and editor of *De Re Metallica,* the Cornish word "stamper," identifying these mills, is Germanic, from *stampfer,* "or, as it is often written in old German works, *stamper.*"[9] At any rate, the German origin of the stamp mill was forgotten, and the device became the Cornish "gravitation" stamp mill, the work horse of English tin mining.[10]

An old Cornish 16-stamp mill, with square wooden stamps, shod with cast-iron shoes, called wearing-members. These shoes were about seven by ten inches, fastened directly to the wooden stems.

A twenty-foot overshot water wheel, mounted on the cam shaft, supplied the power. This particular mill ran at twenty-three drops per minute and crushed about three-quarters of a ton of tin ore every twenty-four hours. (*Mining and Scientific Press*, June 12, 1915)

The Cornish were good miners, but poor mechanics. In nearly 500 years they failed to perfect their "gravitation stamps" much beyond the original German prototype. Yes, they added steam power, but reluctantly. Up to the 20th Century most mills were still driven by overshot water wheels.

Like their German counterpart, Cornish mills remained crude, heavy, and noisy, components changing hardly at all. The 600 pound stamp was comprised of a rectangular mass of iron, six by twelve inches, cast on the end of a twelve foot long wooden or wrought iron stem, or "lifter." The stamp-lifter assemblies stood very close together in the mortar with flat faces towards one another. When the head wore away the stamp was sent to the foundry to have a new one cast on. The stamps did not revolve. For a tappet, or "tongue," a massive iron ring was slipped over the stem and held in place by an iron wedge. These were actuated by a twenty- to twenty-four-inch diameter cam shaft, or "axle," made of hollow cast-iron sections bolted together by means of flanges. Heavy cast heads, eight to ten inches in diameter, bolted to the ends of the "axles," served as journals. The cam shaft was provided with openings, symmetrically spaced, into which steel or wrought iron pieces were inserted and held tight by iron wedges. As the "axle" revolved these projections acted as cams, engaging the tappets, lifting the stamps in proper rotation. One turn of the cam shaft raised each stamp five times. Each cam shaft was long enough to actuate three or four batteries of four stamps each. The speed of Cornish stamps was slow, about twenty-three drops per minute, which gave a crushing rate of about three-quarters of a ton of ore every twenty-four hours.

The mortars, or "coffers," were of wood lined with iron, bedded in earth or gravel, the ore to be crushed

A closeup of the above Cornish stamp mill, showing details of the twenty-inch diameter cam shaft constructed of hollow sections of cast-iron bolted together by means of flanges. There were bearings at each end of the shaft, and between each set of four-stamps. The stamps were actuated by forged iron cams placed in holes casted in the cam shaft. Each stamp had four to six cams.

Bottomless mortar-boxes were bedded in earth or gravel, the ore to be crushed forming the bottom. Therefore, the only wearing members were the cast-iron heads or shoes.

forming the bottom. No dies were used, the stamps simply dropped on a thick bed of ore that always filled the bottom of the "coffer." Instead of ore-feeders, each mortar was provided with a board ore chute extending its whole length. A bin kept this chute supplied with ore which always slid into the mortar as the stamps beat out the bottom.

The whole contrivance worked much like an an old style music box with its cylinders of many projecting pins. Only it functioned on a gigantic scale. Considering that the stamps worked in individual cast iron guides, generally loose and rattling, it may be imagined that the music of a 20-stamp mill was titantic. The noise has been likened to that of "truck loads of iron nails being hauled along a cobbled street."[11]

Cornish and Welsh miners and millmen carried the knowledge of this contraption to North America. In the early 1830s gravity stamps were put to use processing gold-bearing rock in Georgia. The mills were subsequently utilized on auriferous deposits in Tennessee, Virginia, Alabama, and the Carolinas. For two decades Cornish stamps pounded Southern ore with no modification beyond the addition of a copper lined mortar for amalgamation. It is strange that the mill did not undergo further modification as it did in the years following the California gold rush. Perhaps innovation was squelched, as some writers assert, by factors of geography and economics. At that time the south Atlan-

A mortar used in a stamp mill at the Columbia Gold Mine in McDuffie County, Georgia, in 1833. Gold was discovered at this place in 1829; and the mine and mill worked until 1861. (*Mining and Scientific Press*, July 24, 1915)

Gravity stamp mills spread throughout Europe and the British Isles during the Middle Ages. Cornish, Welsh, and German millmen brought them to the New World in the 17th and 18th centuries. To the right is an old stamp mill built in one of the out-of-the-way districts of Oaxaca, Mexico. A caption accompanying this photo states, that "it stands as a monument to the millman of yesterday, who bv axe, timber, and grit fashioned this crude mortar and added another iota of bullion to the world's store." (*Mining and Scientific Press*, September 20, 1913)

tic region was divided into large estates, worked by slave labor. Gold mining was only an adjunct of agriculture, and mechanical skill was not wasted on problems of milling gold ore. Then too, most of these mills were destroyed by the Civil War, their iron components going into Confederate armament. In the case of the southern stamp mill, war may have halted innovation.

Such was not the case in California. Within two years of discovery of gold, 100,000 people had flocked to the Pacific Coast, among them alert, capable, ingenious men who had mining and milling experience. According to Courteney De Kalb they were "fitted for the frontier not only by physical vigor, but by the genius for adaptation."[12] They would transform a Renaissance rock crusher into what history has come to recognize as the California Quartz Mill, a device adaptable to a multitude of mining situations; a device that would make San Francisco the center for the manufacture of mining equipment. For the next fifty years the world would look to California as the source of mineral processing technology.

San Francisco would have become a great port even without the gold rush, for the simple reason that it has a bay of magnificent proportions. It is a body of water up to thirteen miles wide, extending nearly forty miles southeastward and ten miles northeastward. The bay is completely shut in by coastal mountains, except at the Golden Gate. Its 100 miles of shoreline embrace 450 square miles of water with depths up to thirty-six fathoms — a harbor unequaled anywhere in the world. Added to that, the bay is linked to two interior valleys by the American and the San Joaquin rivers, streams of considerable size. Thus it was natural that the sudden unfolding of wealth in 1849 would transform a nearly deserted town into a center of maritime trade. That transformation was not easily made.

Situated on the margin of the North American continent, San Francisco at that time was 2,500 miles west of the nearest manufacturing town of any importance in the United States. It was separated from the great marts of the world by 15,000 miles of sea. Food, raw materials for manufacturing, tools and precision instruments needed for commerce, had to be purchased in New York, Philadelphia, Boston, and Pittsburgh, then brought around Cape Horn or across the Isthmus of Panama and shipped over the Pacific Ocean to the Golden Gate, a voyage of at least 135 days. All this involved loss of time and high freight rates. Until transcontinental railroads were built San Francisco groaned under the handicap of isolation.

The challenge of long distances and high priced imported goods was met by men who did not find El Dorado by mining. Aboard the hundreds of vessels entering the Golden Gate were men, both Americans and Europeans, who possessed professional and mechanical skills. Many worked for shipping lines as rope and cordage makers, blacksmiths, boilermakers, and engine men. Most never sensed the opportunity for profit until they arrived in San Francisco. On the other hand, there were some like Henry Stow, owner of Chicago's first iron foundry, who knew the minute he read of California's gold rush that the region would cry for repair facilities.

That story began in 1809 with the birth of Henry's brother, William H. Stow, at Log City (Hamilton), New York. Not much is known of William's early childhood other than he served his apprenticeship in an iron foundry at Syracuse, New York, becoming a molder by trade. From Syracuse he went to Buffalo where he worked at his trade before heading west to Chicago in 1834. There, he applied his skills.

On Polk Street near the river, he established his foundry, William Stow & Company. With motive power supplied by an Avery rotary engine brought from Syracuse, William built Chicago's first steam engine, and poured out a steady stream of threshing machines, plows, steam drills and pumps, the latter in demand for construction of the Illinois and Michigan canal.

About 1845 or '46 the foundry passed into the hands of William's brother, Henry M. Stow, who was not only a capable mechanic but an opportunist. Smitten with dreams of El Dorado in 1849, he put all the company's machinery, fixtures, tools, coal, pig and wrought iron in stock at the time, on board a vessel purchased and loaded by R. K. Swift of Chicago, bound for California. Shipped as ballast, the foundry was nearly a year reaching its destination, the vessel having to pass into the Atlantic Ocean by way of the Great Lakes, the Welland canal, and the St. Lawrence River. It reached San Francisco after sailing out of the Azores, through the Straits of Magellan, and over a long stretch of the Pacific Ocean.

In 1850 Henry M. Stow landed his foundry on the beach at San Francisco, and moved it a distance of 600 feet. The manhandling of his equipment and supplies cost ten times as much as shipping it from Chicago to that point. In May of that year, Henry removed his foundry to Sacramento where, in partnership with H. A. Bigelow, that city's first elected mayor, he again put

An 1880s view of San Francisco Bay looking toward Oakland from Telegraph Hill. This view includes some points of interest connected with the city's commerce and passenger traffic. Out-going and incoming ships seen here are lying in the channel which constituted the harbor proper. Across this expanse of water plied vessels which went up and down the coast and across the Pacific Ocean. The large island slightly to left center is Goat Island. Just to the right of it, and about a mile to the east, are wharves of the Central Pacific railway. The foundry or Iron District, bordered the waterfront. (*Mining and Scientific Press*, March 5, 1881)

it in operation at the junction of the Sacramento and American rivers. The foundry produced castings, mainly for ship repair, for which it charged $2.50 per pound.

As the price of castings in Chicago at that time ranged from two to three cents per pound, it is likely the earning capacity of the old foundry was not diminished by its removal to the Pacific Coast, although wages paid to employees — an ounce of gold, or $16 per day to each man — had a tendency to keep down profits. This historic foundry operated until 1864, when the price of castings dropped to twenty-five cents per pound. Unprofitable at that point, the foundry was abandoned.[13]

The Stow foundry, Chicago's first iron works, narrowly missed being the first in California. That distinction goes to Michael, Peter, and James Donahue, three brothers from Glasgow, Scotland. In 1849 these young blacksmiths joined the rush to California. Failing to make it washing gold, they assembled a few inferior tools and erected a small crude furnace near the beach in the sand hills of San Francisco. Sheltered in no more than a tent, unprotected from storm, with a furnace blast produced by two blacksmith bellows, they poured the first iron casting turned out on the Pacific Coast.

The Donahue brothers and Henry M. Stow were not the only ones to open foundries. Other men saw the same opportunity. George K. Gluyas planted himself in San Francisco in 1849 and a year later established the Sutter Iron Works on the corner of Beale and Folsom streets. Connected with the Aspinwall Steam Transportation Company, he repaired marine machinery and engines.[14] About the same time, the Pacific Foundry and Machine Shop was erected on First Street; followed by the Fulton Iron Works in 1855, and the Miner's Foundry in 1856. Others followed in rapid succession, until by 1860 there were fourteen foundries and machine shops in San Francisco employing 220 men, repairing ships, building boilers, and doing general blacksmithing. That year these shops produced $1,200,000 worth of machinery, much of which were mining appliances.[15]

From their inception, San Francisco's foundries met the demand of California industry by turning out all manner of iron work, repairing and fabricating marine, agricultural, and logging equipment. They produced picks and shovels, pans and buckets, bells and chains; hundreds, if not thousands, of articles from flat ware to ornamental castings. What San Francisco's foundries and machine shops really excelled at was the manufacture of mining equipment. And that was due to a number of reasons.

Far removed from the scene of Pacific Coast mining, Eastern iron works had little, if any, understanding of the needs of mining men or of Western mineralogy. Then too, distance precluded fast response, something mine developers could not tolerate. Contending with ever-changing geological and geochemical conditions, they chafed at the bit to get properties into production. On the other hand, San Francisco's foundries were close, able to meet the demand for peculiar implements and machinery to solve particular problems. Notwithstanding the high cost of labor, shortages of iron and steel (in the early years foundrymen relied on scrap for the source of their metal), and the necessity of importing raw materials and fuels, the foundry business of San Francisco acquired vast importance, not only for California, but for the Western territories, and eventually for the entire world.

San Francisco's foundry business took off with advent of quartz mining in the 1850s. Almost overnight came the demand not only for hundreds of pieces of iron comprising milling and amalgamation equipment, but also for motive power — ponderous water wheels, steam engines and boilers. For the most part filling these orders was easy. Gold mills, like any other mills, were not complicated devices; the intricacies of steam engine and boiler construction well known. Cut off from Eastern suppliers, San Francisco foundrymen became expert at building steam engines, producing any size of powerplant for any need. As mechanics, they could follow plans, build and erect any kind of appliance demanded by the gold mining profession.

The mining of gold-bearing quartz sowed the seeds of West Coast machinery manufacturing and silver mining matured the industry. Discovery of vast bodies of silver sulphurets at Mount Davidson on the eastern slope of the Sierra Nevada called forth the best of Caifornia's inventive genius. The result was production of great quantities of complex, efficient, and reliable machinery that ultimately revolutionized mining technology worldwide. To fully comprehend the rise of San Francisco's machinery manufacturers, and to understand how their equipment evolved, we need some background on the development of the Comstock Lode.[16]

In May 1850 gold was discovered in the sands near the Carson River. It had been washed down by natural agencies, through Gold Canyon, from the decomposed rock of the ledge above. Within a few weeks shallow deposits of the precious metal were traced more than three miles along the bed of the canyon from the river to Devil's Gate — a narrow, rocky cleft, notable to teamsters as the place where "Uncle Jimmy Fair" collected high tolls that could not be reduced because it "was all he had for the education of his poor children."

From 1850 to 1857 a fluctuating band of miners, whose number varied from twenty to 200, washed the Gold Canyon placers with rockers and long-toms, making from $4 to $5 a day each until the deposits were exhausted. In 1857 the Grosch brothers discovered what history records as the Comstock Lode at some point, now unknown, on Gold Hill. But their discovery perished with them, only to be revived early in 1858 when the Placer Quartz Company located several surface veins around Silver City, built an arrastra and worked some of the ore. The returns, however, were so small that they soon abandoned the business.

The following winter James Finney and his partners found gold in crumbling surface rock at Gold Hill, located claims, and worked the auriferous material in rockers with good results. Other claims were speedily taken up, and by spring of 1859 half a dozen arrastras were profitably operating. The following summer arrastras were grinding ore from new discoveries at Virginia City, and the number of primitive mills at Gold Hill and along the Carson River increased dramatically. A few more mills were started on the Carson River.

Up to that time only gold was sought, for no other metals registered to California miners. But assays made in June 1859 by Melville Attwood, an English metallurgist who built one of Grass Valley's earliest mills,[17] showed the Comstock Lode to be a rich auriferous silver deposit. How to treat the associated silver sulphurets now became an important question. Ore samples then being tested in various California metallurgical works came into the possession of Almarin B. Paul, an energetic Nevada City millman who had also participated in Grass Valley milling. Paul not only knew gold milling as practiced in California, he was also versed on the metallurgy of silver. He decided to treat the sulphurets with chemicals of the patio process in pans then used in California gold mines. Tests made in autumn of 1859 convinced him that this treatment could be successfully applied to Comstock ore.

On November 1, 1859, Paul organized the Washoe Gold and Silver Mining Company. At end of May 1860 he began work on a millsite near Devil's Gate in Gold Canyon. Neighborhood miners laughed at Paul and accused him of incompetency. Despite the ridicule, he ordered machinery from the Miner's Foundry of San Francisco on June 7, and a few days later signed contracts to work several tons of ore from Gold Hill claims (at $30 per ton), the contracts being contingent on the completion of his mill within sixty days — a condition only a bold man would accept in those days, when construction material was far away and transportation difficult and uncertain.

Melville Attwood, who built one of Grass Valley's pioneer stamp mills, was born in 1812 at Worcestershire, England. In 1843 he secured from the British government a patent for a process that gave zinc ore commercial value. He came to California in 1852 as an engineer for the Agua Fria Company, and introduced many innovations to California quartz mining and milling. His assay and report on Comstock ore created the Washoe excitement. When he died at his Berkeley home in April 1898, he was eighty-seven years old.

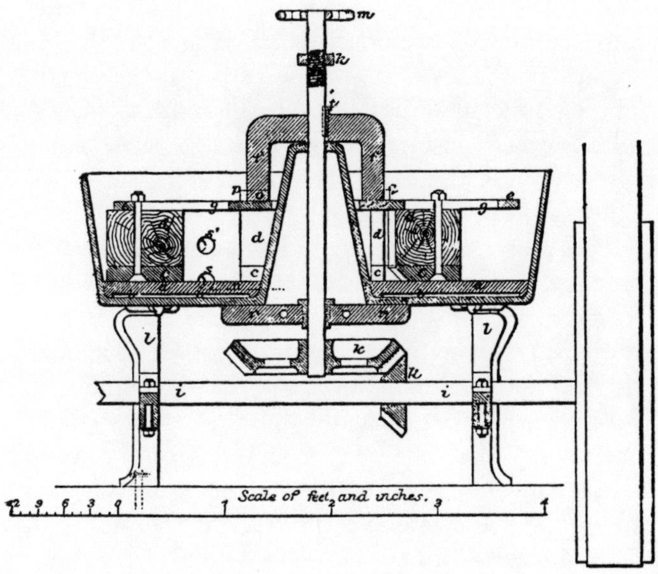

At right: The Knox pan was devised in 1858 by I. W. and R. F. Knox, two brothers who worked as machinists for the Golden State Foundry. This clean-up pan was used to treat battery sands, skimmings, and concentrates, first at Grass Valley than on the Comstock. It consisted of an iron pan, four feet in diameter and fourteen inches deep. Wooden or iron shoes were attached to the arms (g), which made twelve to fourteen revolutions per minute. Iron shoes were considered better for brightening or polishing gold particles contained in pyrites, thus rendering them fit for amalgamation. The pan was charged with about 300 pounds of concentrates, water added, and the pulp ground for three or four hours, after which fifty pounds of mercury was added. The pulp was then mixed for a few more hours. The waste matter was then washed off with a stream of water, and the amalgam retorted. The Knox pan was one of a number of amalgamation devices which were collectively called Washoe pans.

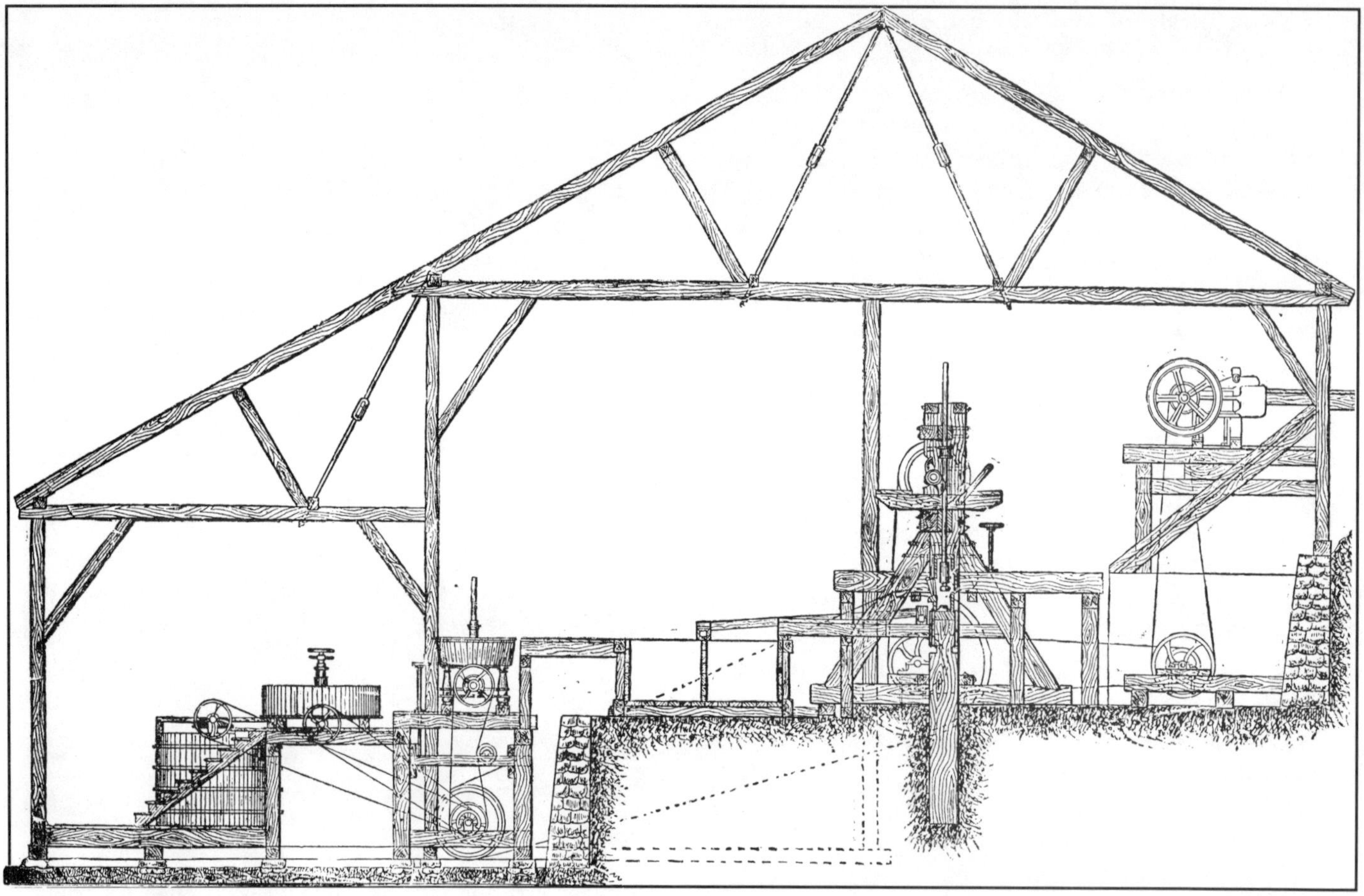

A sectional view of the machinery in a mill working silver ores by the method employed on the Comstock. The batteries of stamps, as many as there may be, were arranged in a straight line. On the feed side was the breaking floor, where rock was prepared by a stone-breaking machine, or by hand, to be fed to the batteries. Crushed ore was discharged upon an apron, or into a trough or launder, which conveyed it to settling tanks, standing directly in front of the batteries. The pulp was moved from tanks to pans where amalgamation took place. From the pans, the pulp was transferred to separators, and agitators, or other similar devices which prevented loss of quicksilver or amalgam. Power to drive this mass of machinery was generated by steam engines and transferred by a maze of shafts and belting. (*Mining and Scientific Press*, December 2, 1882)

Paul was gutsy and determined. He built the mill within limits of his contract, and on August 11 the 24-stamps of the Pioneer Mill were put in operation. Three hours later, Charles S. Coover and E. B. Harris, who had been inspired to rivalry by Paul's example, set in motion the machinery of their little 8-stamp mill built at nearby Gold Hill.

Paul crushed the ore dry in his batteries and amalgamated it in four-foot diameter Knox pans, which held charges of about 300 pounds each. The light iron mullers, revolving twelve or thirteen times per minute, did not grind but served merely to stir the pulp for the purpose of distributing quicksilver throughout the mass. Copper plates fastened in the pans hastened silver reduction. A portion of the amalgam adhering to the plates was scraped off twice a day; the rest drawn off at intervals through a discharge-hole.

Each charge was treated with forty pounds of quicksilver, a pint of salt and a few ounces of copper sulphate, or if that was not available, copper filings. The pans were set on brick furnaces, heated by wood fires, and the water used was warmed by boilers that furnished steam for the mill's engines. At a later date, in place of brick furnaces, iron stoves were used with circular flues, through which the heat and smoke were led beneath the pans. Then Paul suggested steam-chambers fitted to the pan bottoms, a suggestion adopted and perfected by William H. Howland. Thus, Paul invented the Washoe Process, the first successful application to treat silver ores on a large scale by pan amalgamation.

Almarin Paul brought to the Comstock a cheap, fast method for the reduction of silver ore. His process would be improved by others. N. P. Boss introduced the continuous feed of ground ore into connected

A model quartz mill as depicted in the July 23, 1864, issue of the *Mining and Scientific Press.* The upright portion of the mill, in the center of the picture, are the stamps, divided into four batteries of four stamps each. The stamps worked in a long iron mortar into which ore was fed from the front of the battery. The crushed ore passed through screens to the settling vats, a row of which can be seen in front of the batteries. A row of separators sit below the vats. When the quicksilver had settled in the separators, it was drawn off into a suitable vessel by means of the tube shown in the engraving. It was then strained through thick canvas or buckskin, the gold remaining in the strainer in the form of a dry, hard "amalgam." The amalgam then placed in a retort, heated to drive off the mercury, leaving a hard, spongy mass of metal.

pans, as well as an automatic method of handling quicksilver, an important matter when a single mill handled as much as sixty tons of that metal every day. For the next twenty years the Washoe pan would be the chief agent in separating $500,000,000 worth of precious metal from its ore.

Paul's first mill-runs, and those of Coover and Harris, gave satisfactory results, and materially increased the rush to the Comstock. Claims were taken up in the neighborhood of the lode, and a large number of mills were built in 1860. Paul's company (the Washoe Gold and Silver Mining Company, No. 1) at once put up a larger mill at Gold Hill, which sparked a mill building mania. By 1861, seventy-six mills, with 1,152 stamps, forty or fifty arrastras, and several patio-yards, were treating Comstock ore. An additional twenty mills were in the planning stage. Reduction works lined Seven-Mile, Six-Mile and Gold Canyons, from Virginia City to the Carson River, and along the Carson River, from above Empire to below Dayton.

The rush to build Comstock mills taxed San Francisco foundries and machine shops. Two thousand men worked around the clock, seven days a week, designing and perfecting mining and milling equipment. The output was enormous. "Hundreds of creaking wagons, burdened with their freight of iron," commented Elliot Lord, "were dragged over the steep grades of the Sierras to the valleys beyond."[18]

The Washoe pan immediately put the Comstock on a paying basis, resulting in feverish activity. In two years 700 mining and milling companies were incorporated to exploit the lode. Most of the mines never produced anything except litigation and assessments, and a large number of mills were so hastily and poorly built that they were quickly antiquated by the rapid improvement in machinery and technology. Stockholders of companies lucky enough to have survived, demanded more and more production at the least expense. Mine and mill superintendents responded by applying pressure to San Francisco foundries for more efficient equipment. An engineering frenzy ensued.

Stamp mill components were redesigned to accommodate a hard ore whose precious metal content was coarse. Mortars used to dry crush gold ore were altered to crush silver ore, wet or dry. Mortars were enlarged. Circular stamp heads were designed and made to revolve at every lift which resulted in even wear and more effective work. Weight of stamps and their rapidity of drop increased. By trial and error, millmen found that the quickest and most profitable way to free gold and silver was to pulverize the quartz with heavy, quick dropping stamps.[19]

Weight of reciprocating parts (cam, stamps, and shafts) of the California stamp mill were reduced until the weight of entire outfit was a fifth that of Cornish gravity mills. Engines too were altered to work at pressures as low as forty pounds per square inch. As fuel had to be transported over long distances, this was of critical importance.[20] Last but not least, foundrymen eliminated the expense and tedium of hand feeding stamp mills by attaching self-feeders.

When millmen complained of time lost in replacing stamp mill's copper plates due to oxidization of their mercury covered surfaces, a San Francisco brass founder silver plated the copper sheets, improving adherance of mercury and increasing recovery of precious metal. As a result, silver-plated copper plates became a standard item in all stamp mills.

Milling practices were not the only thing the Comstock revolutionized. Its deep and scattered ore fissures altered mining methods. At first there was no such thing as systematic exploration on the Comstock. Mineral-bearing streaks and veins were followed by burrowing and trenching, then by sinking incline shafts; the ore hauled to the surface in ironbound buckets suspended by hemp ropes from man- or horse-powered windlasses and whims. One bucket held no more than 400 pounds. That changed with introduction of steam power.

In 1860 the Ophir Mine was plagued by an increasing flow of water into its 180-foot deep incline shaft. A pump powered by a fifteen horsepower donkey (auxiliary) engine was installed. The pump not only momentarily controlled the flow of water, it provided faster ore haulage. The resourceful superintendent fastened a rope directly to the engine's mainshaft and used it to raise and lower a car through the shaft — an innovation rapidly adopted by other Comstock companies. Within two years twelve companies had steam driven machinery for pumping and hoisting.[21]

Steam-powered haulage would have failed had it not been for the introduction of wire rope. Greater tonnage hauled from great depth called for ropes of high tensile strength. Hemp ropes would not do when it came to lifting tons of ore or lowering men and machinery. Round iron ropes were tried with indifferent success, owing to poor quality of material and faulty design of hoisting reels. In September 1863 the first flat iron-wire rope was made by A. S. Hallidie & Company of San Francisco, for the Sierra Nevada Mine. It was woven in two coils, each 700 feet long, and was four inches in breadth by half an inch in thickness. Warranted to raise ten tons, it was an important improvement upon ropes in ordinary use, as it was less liable to slip on the barrel of the reel and a cage could be lowered more steadily. Similar cables, made of braided steel wire, were subsequently adapted by the Savage, Hale & Norcross, and a majority of other mines on the lode.[22] Steam powered hoists and wire rope, all products of San Francisco manufacturers, set the stage for exploitation of the Comstock at greater depth.

The first line of exploration on the Comstock Lode was relatively shallow, no more than 500 feet. Ore along that line was worked out by 1869, and the focus of attention shifted eastward a thousand feet or more, where a new line of shafts was commenced. Because orebodies were much deeper along that horizon, 1,000 to 1,200 feet, more powerful machinery was required. Fortunately for the companies, in the late 1850s San Francisco foundries had perfected for Grass Valley quartz mines equipment capable of lifting from such depth.

Two classes of hoists were readily available from the major foundries: single and double hoists. Of these, there were again two types — first and second motion engines. In the former, two engines with cranks at right angles to each other were coupled direct to the drum shaft; in the latter, the engine carries a pinon, meshing with a spur-wheel keyed to the drum shaft. The rope drums in either case were proportioned in diameter to the size of rope employed, and in length, to the depth of the shaft. Single compartment shafts called for one drum, but two-compartment shafts had to have a hoisting engine provided with two drums, each having a separate rope.[23]

These hoists were powered by the favorite American steam engine, the Corliss, a powerplant that had been around for nearly twenty years. This engine — the prototype of which was introduced February 17, 1848, by George H. Corliss — was of the walking-beam type, having the regulator combined with the cut-off valves. It usually had a twenty-two-inch diameter cylinder, and a six-foot stroke, and worked under sixty pounds of boiler pressure. It was capable of developing in excess of 250 horsepower.[24]

It took fifteen years for steam-power to evolve on the Comstock, a revolution coming in 1875 with consolidation of major mining properties into the hands of a few San Francisco capitalists, who immediately turned their attention to striking deep ore carrying fissures. A series of shafts serving the needs of several companies were sunk. Designed to reach depths of 4,000 feet or more, these combination shafts as they were called, required the heaviest machinery ever designed on the West Coast, equipment that would not be replaced for years. This time Grass Valley did not furnish the prototype for the equipment. Instead, the new Comstock equipment was patterned after hoists installed on the Mount Diablo coalfield in Contra Costa County.[25]

These shafts were impressive and so was the equipment; engines and hoists exquisitely built. "Intended to a shaft for all time," the great hole which the Savage, Hale & Norcross and the Chollar-Potosi began excavating in 1875 awed Dan De Quille. "It was of vast size, containing several spacious compartments for hoisting and pumping purposes, and was supplied with the most powerful machinery then manufactured." Novel equipment was set up by the Savage Company as it prepared to sink a shaft to a depth of 4,000 feet. According to De Quille, the "novelty" was the reel on which the hoisting cable wound, the first such drum on the Comstock.

"The reel is fifteen feet in length, and at the larger end is twenty-two feet in diameter, while at the smaller end the diameter is but thirteen. It is suspended upon wrought iron shaft about sixteen inches in diameter, the ends of which revolve in ponderous bearings supported by foundations of cut stone reaching into the earth to solid rock. The shell of the reel is covered with thick wooden staves, and the whole somewhat resemble a great tapering cask. Over the staves are securely bolted heavy iron plates forming a strong armor outside of the wooden structure. In this iron armor is a deep groove, which, starting at the smaller end of the great conical drum, runs in a spiral manner to the larger end, just as the groove between the threads of a screw is seen to run. In this groove winds the cable as the incline-car ("giraffe") is let down into or drawn up out of the mine.

"When the car is at the bottom of the incline, the greater part of the cable is off the reel, and when the hoisting begins, it is wound up on the smaller end of the drum, where the engines have greater purchase on the load. As the hoisting proceeds and the weight to be raised becomes momentarily lighter, on account of the heavy steel cable being wound up, the lifting force is steadily moved toward the larger end of the drum, and each revolution adds to the swiftness of the ascent of the car that is being raised. The cable is round and is made of the best steel wire. It is 4,000 feet in length, and weighs 25,190 pounds. The upper part, for a distance of 1,500 feet down, is two inches in diameter; for the remainder of its length, 2,000 feet, it gradually tapers till at the lower end its diameter is one and three quarters inches. The taper is not made by dropping wires in the several strands of the rope, but by drawing each wire (as it is manufactured) slightly tapering for the lst 2,500 feet of its length."[26]

The excellence of Comstock hoisting machinery was due to the expertise of San Francisco foundrymen, who carefully listened to the needs of such Comstock engineers as Behr and William H. Patton, and crafted complex machinery that functioned exceedingly well. By 1876 rock was being raised from a depth of 3,000 feet at speeds approaching a mile a minute. Considering that most of this equipment was put together on production schedules of ninety days or less, that was an amazing feat.

As mines went deeper, temperature within the workings went higher, ventilation became poorer. Hot, foul air debilitated miners. Downcast shafts were sunk to suck in fresh air. Roots blowers were ordered from Sacramento and installed in the mines to keep stopes and drifts cool and free of foul air. Compressed air brought even more relief. On top of that, mines were becoming exceedingly wet.

Until 1863 small pumping plants were able to keep the Comstock comparatively free of water. Then on January 18, 1863, a miner in the Ophir Mine sank his pick into a clay seam in a cross-cut 313 feet below the surface. The sprout of water that followed drove men from the workings. Other companies were plagued by flooding: Best & Belcher, the Crown Point, and the Overman Mining Company. More pumps were put in operation and drainage shafts were cut. By December 1864 the Belcher Mine was discharging over a million gallons of water every twenty-four hours. Much of this water was scalding. Approaching temperatures of 160 degrees, this water added to the ventilation problem, and threatened production. It was misfortunate in one sense, a blessing in another way.

This tide of water spelled profit for those equipment manufacturers who could fabricate engines of Cornish design, capable of lifting columns of water

Shaft sinking and drifting with machine drills. These Ingersoll drills are arranged for vertical and angular boring in the shaft bottom, and drills mounted on columns are working in drifts extending each way from the shaft. A bucket is being lowered with a new set of drills for the machines. The drills were powered by compressed air ducted by pipe throughout the mine. (*Mining and Scientific Press*, May 5, 1883)

from a depth of nearly half a mile. Of the fourteen foundries in San Francisco in 1865, only three had the expertise and equipment to build Cornish pumps — Union, Pacific, and Risdon iron works. The amount of hydraulic equipment these three foundries turned out was staggering; a combined pumping plant, with nearly three miles of massive pump rods, balance-bobs, pumps, and moving parts. In all some 6,496,400 pounds of iron, steel, and wood, costing $1,300,000, not including freight, foundations, and installation. Even with installation of such a massive system, only 5,400 gallons per minute could be raised; the contest against water was a losing fight.[27] Nevertheless, flooding of the Comstock wrought great improvements in pumping gear, much to the benefit of the mining industry. Even the introduction of the pneumatic rock drill to Western mining was the result of the fight to control water on the Comstock.

What was needed was a cheap, effective method of drainage, something like a tunnel extending from the foothills of the Carson Valley to intersect the Comstock Lode. Men moved to turn that dream into reality. On February 4, 1865, the Sutro Tunnel Company was incorporated to drive an adit more than three miles to a point 1,663 feet below the top of the nearest shaft, the Savage Mining Company. Estimated to cost $3,000,000, the tunnel would be paid for by the mining companies who pledged two dollars for every ton of ore extracted from their respective mines after the extension of the tunnel and its lateral drifts to designated points within their boundaries. Thus, the Sutro Tunnel would not only serve as a drainway, but also an avenue for transportation of ore, men, and equipment. Water flowing through the adit would turn turbines, furnishing electricity to power mills and mines. A bold plan indeed.

Work on the Sutro Tunnel commenced in summer of 1868. It was slow, hard labor, accomplished by hand-drilling and blasting. By 1873 only 1,919 feet of adit had been cut. At that rate it would take over seven years to complete the project. That would not do; a faster method of drifting had to be found. The solution was found in a pneumatic drill and compressor perfected by Charles Burleigh.

The Burleigh drill was created at the Putnam Machine Works in Fitchburg, Massachusetts, a foundry in which Charles Burleigh had a partial interest. His device, however, was not entirely new. The idea of a mechanical drilling machine followed close upon the heels of the 1613 introduction of a method of drilling and blasting proposed by Martin Weizel of Freiberg, Germany. In 1683 Henry Huthmann of Stefeld advanced the idea of a drilling machine based on the principle of a ram. Others dabbled with the idea for the next two centuries; Gainschnigg in 1803, Richard Trevithick in 1813. No practical results came of these ideas until 1844. That year Englishman John D. Brunton proposed the use of compressed air to work a hammer, striking at the rate of 200 blows a minute.

Others saw merit in Brunton's idea. In 1851 a Frenchman named Cave constructed a drilling machine worked by compressed air. A scheme to tunnel through Mont Cenis on the border between France and Italy spurred development of pneumatic drilling. When word of that project hit numerous inventors stepped forward to offer their drilling machines: Bartlett of England in 1854, Colladon of Switzerland and Schuhmann of Germany in 1855; Castelain and Schwarskopil in 1857, and Sommeiller, Grattoni, Grandis, and Haupt in 1858. By then patent offices worldwide were flooded with schematics of drilling devices, 243 patents alone being granted in the United

States. With few exceptions, notably the diamond drill, all these machines were so-called percussion drills.

It would take another tunnel project to perfect the pneumatic drill — the great Hoosac railroad tunnel through the Green Mountain range of Massachusetts. This tunnel was a project worthy of American ingenuity. It was the longest railroad tunnel in the United States, four and three-quarters miles. Begun in 1854, it took twenty-one years to complete, and cost $10,000,000. A reliable pneumatic percussion drill was the outgrowth of that contest, but that product came slowly, developed through trial and error.[28]

All devices used at Mont Cenis were tried at the Hoosac tunnel. Only the Sommeiller machine came closest to meeting the requirements of American engineering. In the end it too was discarded. The most practical drill was one introduced by Brooks, Burleigh & Gates, ten of which were put to work in sinking the central shaft. Thirty-nine more were employed in the headers. Breakdowns and costly repairs of the machines, however, led to abandonment of the project in 1862. That did not stop Burleigh and his colleagues. They redesigned their drill and compressor, and in 1866 a greatly improved drill was put to work. More substantial, the machine worked on an average of five days without repair.[29]

Equipped with improved Burleigh machines, the drilling crews completed the Hoosac tunnel in 1875. In the meantime, the drills were tested at other tunneling projects: the Nesquehoning tunnel in Pennsylvania, the Lewis tunnel in Virginia, and several tunnels in Colorado. Meanwhile, Benjamin Thomas Lacy, a machinist who helped Burleigh improve his drill, drove thirteen tunnels under Hell Gate reef in New York Harbor. The drifts were packed with explosives and the barrier to shipping eliminated.[30]

With such success, the American mining community was soon clamoring for Burleigh drills, and by 1872 the machines were at work in the copper mining region of northern Michigan. From there it was an easy hop into the Western mining market. Acting as Burleigh's representative to Western America's mining region, Benjamin Lacy introduced the drill and compressor to California quartz mining in 1872, and a year later he opened an office at Virginia City as Burleigh's representative on the Comstock.[31] It was not long before this machine drill was giving satisfactory results in the Belcher, Gould & Curry, Consolidated Virginia, and Overman mines, much to the chigrin of San Francisco machinery manufacturers, who claimed "Mr. Burleigh's drill was born on the wrong side of the mountains."[32]

As it was, the Burleigh drill was put to good use. The machine performed exceptionally well in deepening shafts, and under supervision of Lyman Curtis Parke, an expert in compressed air applications from Pontiac, Michigan, Burleigh drills speeded the driving of the Sutro Tunnel. Completed in 1878, the bore produced immediate effects. The Sutro Tunnel created a new and lower working surface for Comstock mines, making it possible to sink to the 3,000 foot level with the same facility as was formerly done from the surface. In short, operating depth was reduced about 1,500 feet.

The Sutro Tunnel altered the entire hydraulic system as originally designed and installed on the Comstock. Costly, massive pumping engines were left on the surface, while all intervening pumps were removed to below the tunnel-level, where they were doubled up, two being placed at each tank-station. This change left about 1,700 feet of heavy pump-rods between the surface engine and the first set of pumps. The equipment was required to perform double the work load. The result was fatigue and breakage of rods, balance-bobs, pins, and strapping plates; an open invitation to disaster.

And that disaster came on February 13, 1882, when a heavy flow of exceedingly hot water from the 2,810-foot level of the Exchequer Mine flooded the Comstock's lower levels. Despite all pumps running to full capacity, the flow of water could not be controlled. On March 28, 1882, pumping was halted and the water rose to within a few feet of the Sutro Tunnel level, idling the principal mines for the next sixteen years.

Technology developed the Comstock, but technology could not at that time keep up with demands of the lode. Nevertheless, this great auriferous silver deposit left a legacy — it served as a beacon, guiding the mining world. Experience gained in working the lode revolutionized mining. Engineers and workmen skilled or schooled in methods of the Comstock spread their knowledge to every quarter of the globe. The great lode pioneered the way for others to follow.[33] That was only part of the legacy. As Dan De Quille observed, "the mines of the Comstock gave life to the whole Pacific coast;" they provided bread and butter, board and room to thousands of skilled men beyond Virginia City. The Comstock was the mainspring of all kinds of West Coast trades and businesses.

As previously mentioned, at 1860 San Francisco had fourteen foundries and machine shops, producing $1,000,000 worth of goods and services. Stimulated by Washoe mineral production, the number climbed in two years to fifteen foundries, three boiler shops,

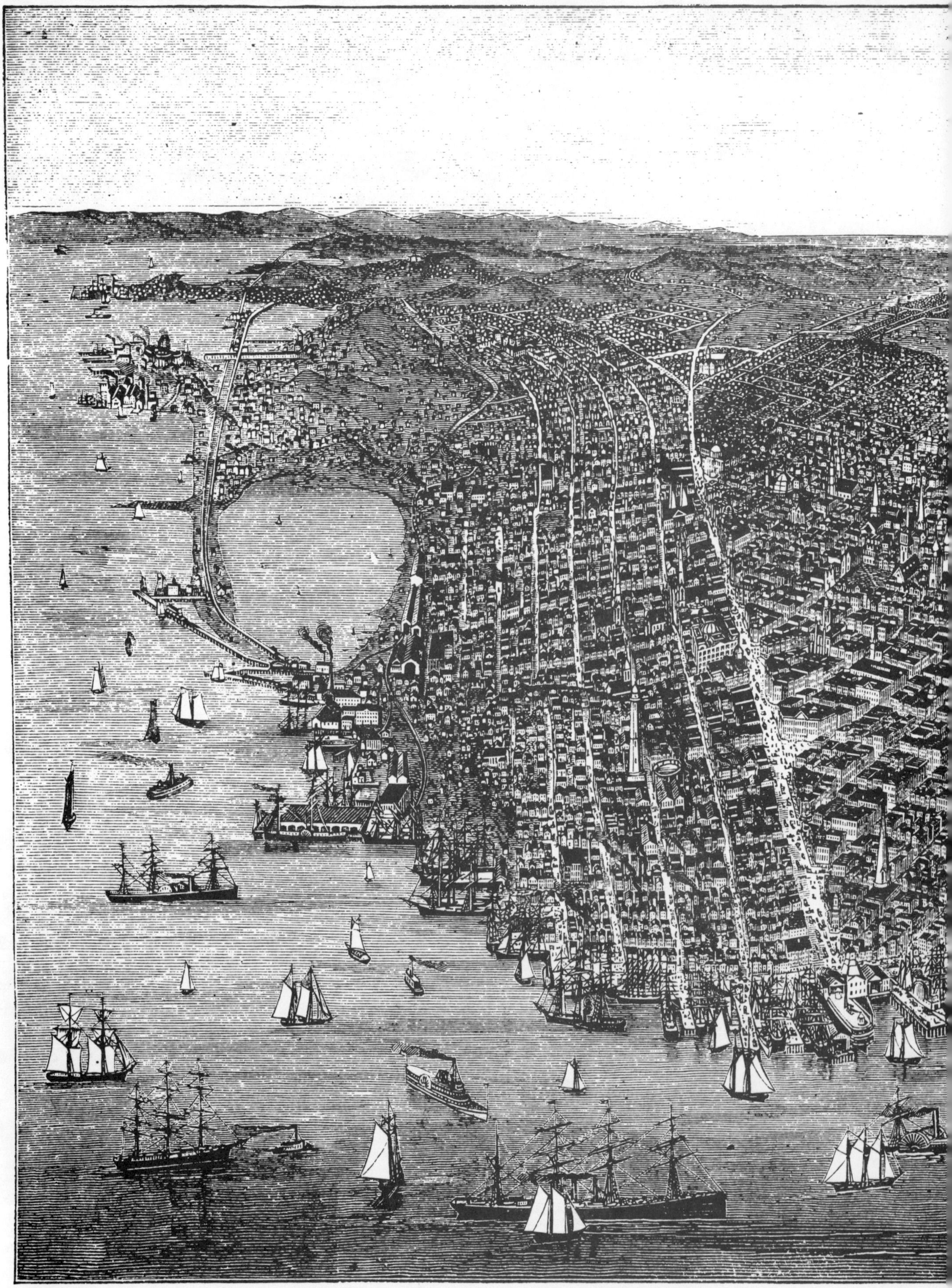

San Francisco's waterfront during the 1870s showing Market Street, the main thoroughfare through town (left of center). Most of the foundries were located in the Iron District close to the waterfront on Mission, Howard, Folsom and Harrison, parallel streets to left of Market Street. The smokestacks of the Union Iron Works can be seen at left center. Above and to the left, marked also by smokestacks, is the Pacific Rolling Mill, located on the Potrero.

and eighteen machine shops, grossing $3,000,000. Although the number of combined shops stood at forty-seven in 1864, with a gross product of $4,610,000,[34] production of woolen goods was San Francisco's leading industry. Comstock development would change that.

By 1876 demand for mining appliances had altered both the economy and face of San Francisco. At great expense, seawalls were erected in the cove, and the sand hills transferred behind the barriers.[35] On this reclaimed land grew the Iron District, where sound of metal turning machinery, the roar of furnaces and hearths, the clank and bang of smiths resounded day and night, six days a week. By 1876 nearly 2,000 men and boys, two thirds of whom were skilled craftsmen, had pushed iron work, most of which involved the manufacture of mining equipment, to the forefront of San Francisco's industry.

Interestingly, San Francisco's number one industry was not built by American citizens. The city's iron trade reflected the tide of humanity drawn to California by the Gold Rush. Most employees came from England, Germany, and the Scandinavian countries. Chinese were excluded from the trade. Considering the times, they made good money. Foundrymen commanded from $3.75 to $4 for a ten hour day. Extra skillful molders got $5, brass workers and cupola tenders, $3 for an eleven-hour shift; finishers, $3 for ten hours. Model or pattern makers, $4 per day; when working by the job or piece, from $5 to $7.50 per day. Blacksmiths, $3 to $4 per day; and boilermakers generally received $3.75 per ten-hour day.

The wealth of the Comstock transformed the foundry and machine shop. A quarter of the forty-seven companies became corporate entities with design and production facilities. Lucrative, on-going contracts for machinery abolished the blacksmith persona; men who had been brawny, grimy smiths and boilermakers became consultants to milling and mining companies. If they could not answer questions pertaining to geochemistry, they hired men who could. Melville Attwood, Almarin B. Paul, William P. Blake, John Hays Hammond, and many others advised San Francisco manufacturers. The transformation went further than that.

As foundries and machine shops became full-fledged manufacturers, customer service became a major issue. Technicians had to be based in mining districts to install equipment and keep it functioning, the result were branch foundries and machine shops, such as the Fulton Foundry in Virginia City, a subsidiary of the Fulton Iron Works of San Francisco. The same company also established a subsidiary in Tombstone, known as McCone and McAllister. Such service was naturally part of promotion and salesmanship.

By the mid-1870s all San Francisco equipment manufacturers had knowledgeable salesmen ranging North, Central, and South America, soliciting orders. The companies lobbyed in every country having any kind of mining industry. Tactics they employed in securing contracts were not unlike those used in today's aviation industry. Politicians, diplomats, and industrialists were wined and dined; their lusts satisfied. Nothing was spared when it came to selling mining appliances. In support of their salesmen, the foundries had well organized programs advertising their products.

Product catalogs made their appearance in the mid-1860s. Now collectors' items demanding high prices, these promotional pieces, illustrated with line engravings, explained in minute detail the workings of mining equipment. Nearly all of the catalogs, issued annually, contained useful information on general mining, ore dressing and reduction. Besides advertising brochures, manufacturers relied on two other means to spread knowledge of their products — an annual trade show and a trade paper. Although both paralleled foundry growth, the trade show appeared first, the outgrowth of an organization of mechanics.

On December 7, 1854, Bay City mechanics, foundrymen, and tradesmen met at the office of the city's tax collector to organize an association that would promote the "welfare and advancement of the workingman and of home industry throughout the State."[36] With George K. Gluyas presiding, and Roderick Matheson acting as secretary, the tradesmen voted to organize the Mechanics' Institute of the City of San Francisco. Benjamin Haywood, Roderick Matheson, J. S. Williams, E. T. Steen, and Henry Carlton drafted a constitution and by-laws; and by March 29, 1855, the Institute was fully organized with 282 members.

The Institute's first slate of officers were: president, Benjamin Haywood; vice president, P. B. Dexter; recording secretary, William M. LaRoche. A board of directors consisted of Gardner Elliott, George Coffran, S. H. Williams, James Ballentine, John C. Macready, Eli Cook, and S. G. Bugbee.

In early April 1855 the Institute established a library and readingroom, a far reaching decision, for West Coast industry cried for technological information. S. G. Bugbee placed the first book on its shelf, Louden's *Encyclopedia of Architecture*, and within a short time the repository had 487 volumes. In 1857 another 400 volumes were added, mostly donated by

The Mechanics' Pavilion of 1858

The Mechanics' Pavilion of 1864

The 1860 Mechanics' Pavilion

Samuel Brannan, Thomas O. Larkin, and James Lick.

Although the library got off to a slow start, the Institute being at first cash strapped, most industrialists recognized the need for an up-to-date reference and technical library to support their inventiveness and manufacturing. Some like Andrew S. Hallidie, who in March 1869 was elected president of the board of trustees of the Mechanics' Institute, envisioned the library being patronized by "the unemployed and roving youth of the city . . . to prepare themselves for the mechanical trades," there being at that time no vocational or industrial schools for training artisians in San Francisco.[37]

It was a Pennsylvanian of German ancestory who really pumped life into the Mechanics' Institute. A carpenter, joiner, and piano-maker, James Lick arrived in San Francisco in 1848, on the eve of the Gold Rush. With $30,000 obtained somehow in South America, Lick began buying real estate in California. He acquired land in the Santa Clara Valley and on the shores of Lake Tahoe, built a flour mill in San Jose and a hotel in San Francisco. He even owned a portion of Catalina Island, and later acquired land at Virginia City. He made an enormous amount of money which he was rather generous with. Lick was there to help the Mechanics' Institute.

In 1857, with only $300 in its treasury, the Institute envisioned an industrial exhibit to promote California industry and attract members. How to finance the venture and where to hold it was the organization's delimma. Lick came to the rescue with both money and land. He endowed the Institute with $400,000 and tendered the free use of property on Montgomery Street, where he envisioned building a home. A 20,000 square foot pavilion was erected, designed in the shape of a Greek Cross, surmounted by a large swelling dome, ninety feet in diameter. On September 7, 1857, the First Exhibition of the Mechanics' Institute opened. For ten days crowds inspected 941 industrial displays. So successful was the exhibition that the pavilion was left standing for a Second Exhibit which commenced September 1 the following year. That show went on for fifteen days, setting a pattern for subsequent fairs.

Exhibitions followed at yearly intervals; the scope of the show growing. By 1864 it had six departments: raw materials, machinery, manufactures, natural production, works of art, and miscellaneous. The best exhibits in each category were awarded prizes, and publicized in California newspapers. The Pavilion was the natural place to introduce machinery, and manufacturers seized upon the first three categories. This industrial fair became the only trade show in Western North America where miners and millmen could view the latest appliances, in most cases running.

A good example of what the Pavilion offered is found in the 1864 list of exhibitors. The Pacific Foundry displayed a fifty-horsepower steam engine, working with Green's Patent cut-off. "The engine was finely finished and is a most creditable piece of workmanship," reported the *Mining and Scientific Press.* Quartz mills was the largest, and noisiest, category on exhibit, with a Pacific Foundry Bryan five-stamp battery, a Howell Rotary Crusher from the Vulcan

Alfred T. Dewey

Warren B. Ewer

Foundry, and a Golden State Foundry stamp mill. A Blake Stone Breaker, an Eastern invention designed to prepare quartz for stamping, was also demonstrated for the benefit of California's millmen. And there was an array of grinders and amalgamators, as well as furnaces, retorts, hydraulic equipment, the latter including pumps, nozzles, and water injectors.

Mining equipment grew in variety and complexity with each trade show, and that equipment and its show received increasing press coverage after 1860. The publicity drew mining men from every district in North and South America, as well as industrialists from further afield; Russians, Japanese, Chinese, and Australians. In many cases, these men purchased equipment from the floor of the Pavilion. That press coverage was provided by the *Mining and Scientific Press*. Without that paper the annual Mechanics' Pavilion would not have had the impact it did. The *Mining and Scientific Press* furnished the advertising that turned the trade show into a yearly extravaganza. The two worked hand-in-hand to the great benefit of West Coast industry. At this point we need to explore the background of this journal.

The *Mining and Scientific Press* had a humble and somewhat shaky beginning. Founded in San Francisco by George H. Winslow and Julius Silversmith, the first issue (an eight pager) appearing on May 24, 1860, under the masthead of *The Scientific Press, "A Journal of Science, Art, Manufactures, Chemistry, Invention, etc."* Winslow did not live long enough to see his paper germinate. Less than a month later he drowned while trying to rescue a child who had fallen into San Francisco Bay. That tragedy dissolved the partnership and Julius Silversmith assumed full editorship, announcing that the paper would "give every week a treatise upon all the arts, science, trades, and manufactures, which will not only be entertaining but vastly useful. These articles will be written by the most talented in this State, who have kindly volunteered, and whose literary acquirements stand preeminent, combined with a practical and theoretical knowledge of the subjects they may treat of, which will not fail to create the most intense desire for every issue."[38] Silversmith intended doing all this by placing editorship of the paper in the hands of "an association of scientific men." A tall order indeed.

Right from the beginning Silversmith focused on the Mechanics' Pavilion, publishing daily issues during the duration of the show. On September 1, 1860, he announced that because of California's growing

At right in descending order: The home of the *Mining and Scientific Press*. The Old Capitol Building, 1862 to 1863. Building at 202 Sansome Street, the office in 1877. Interior of office at 202 Sansome Street. The last illustration, the building at 330 Market Street, office from 1897 to 1906. This building was destroyed by the earthquake and fire. (*Mining and Scientific Press*, May 21, 1910)

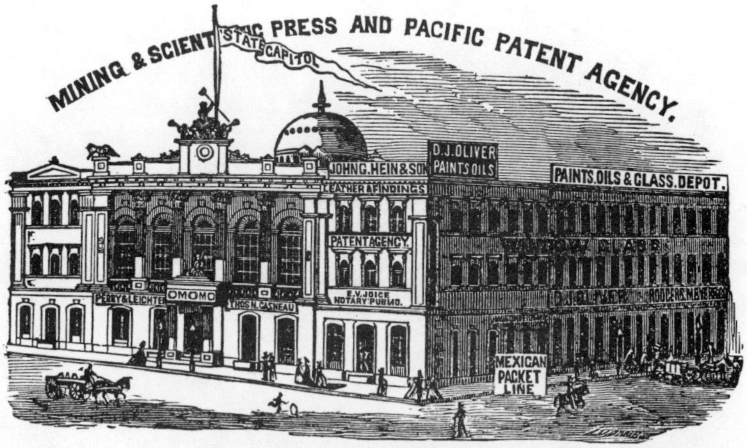

mineral wealth, mining would thereafter be the chief feature of the paper. To reflect that interest, on October 20, 1860, he changed the paper's name to the *Mining and Scientific Press and Mining Advocate.* From that date on the paper steadily increased in size, the bulk coming from numerous articles on mining, metallurgy and kindred subjects.[39]

The *Mining and Scientific Press* struggled to fulfill its pledge. "The association of scientific men" soon tired of editing the paper, for the November 30, 1860, issue bore Silversmith's name as both publisher and editor, and continued so until October 7, 1862. Nevertheless, he accomplished his primary goal, filling the paper's columns with mining news. At the same time he began covering San Francisco's industries, including the emerging foundry business. Without help, putting together a trade paper was hard, lonely, and an often thankless task. Fatigue apparently set in and Silversmith bailed out.

On November 1, 1862, he was succeeded by W. B. Ewers, who was joined by C. W. Smith on January 19, 1863. They shortened the publication's name to the *Mining and Scientific Press.* Under their energetic management the paper became the recognized authority on mining and metalurgy. Nearly every step in the invention of mining equipment, or improvements of metallurgical processes received first notice in its columns.

A year later, on December 14, 1863, twenty-nine year old Alfred T. Dewey acquired a half interest in the paper. A New Englander, Dewey learned the printing trade at Westfield and Springfield, Massachusetts, where he and his brother published both weekly and daily newspapers. In 1856 Alfred came to California and purchased a half interest in the La Porte *Mountain Messenger*, a paper he later became sole owner of.[40] On December 14, 1863, Dewey entered into a

Almon D. Hodges, Jr., editor 1869-1871.

Charles G. Yale, editor 1871-1894.

partnership with Ewers and Smith to publish the *Mining and Scientific Press,* the firm being known as Dewey and Company. This partnership continued until August 8, 1868, when Smith sold his interest to Dewey and Ewer, who continued as partners until the Dewey Publishing Company was incorporated in December 1891.[41]

According to H. F. Dexter, Dewey's nephew, "the early owners of the paper seem to have been of a migratory nature; possibly they found it 'cheaper to move than pay rent.' The original office was in the Government House, corner of Sansome and Washington streets. On January 11, 1862, the office was moved to 522 Merchant Street; on April 9, 1862, to the State Capitol building, or as it was afterward designated, the Federal building, corner of Battery and Washington streets; on September 21, 1863, it was moved back to the Government House; on April 10, 1865, to 505 Clay Street; on August 8, 1868, to 414 Clay Street; on October 1, 1871, to 338 Montgomery Street, southeast corner of California Street; on July 13, 1874, to 224 Sansome Street, corner of California Street; and on July 28, 1877, to 202 Sansome Street, northeast corner of Pine Street."[42]

Dewey and Company had many irons in the fire. Besides publishing the *Mining and Scientific Press,* it produced the *Mechanics' Fair Daily,* consisting mainly of write-ups of exhibits during the duration of the fair. In 1871 Dewey established the *Pacific Rural Press* which gave a voice to the State's agricultural community. Dewey and Ewers, with help from George H. Strong, also functioned as patent agents. Their American and Foreign Patent Agency for the West Coast solicited letters of patent in the United States, British Columbia, Europe, "and other civilized countries," for "inventors and discoverers in arts, mechanics, metallurgy and chemistry." Most mining appliances and mineral reduction processes developed in California prior to 1890 were patented through Dewey and Company.

To the great benefit of the mining industry, Dewey and Company were also job printers who pioneered West Coast typography and engraving. The *Mining and Scientific Press* was set with type produced by the Faulkner Type Foundry, San Francisco's first type foundry, and printed on a Chicago Taylor Combina-

J. F. Halloran, publisher 1894-1906.

T. A. Rickard, editor 1906-1909.

tion Press. Upon this press, acquired in 1873, Dewey printed most of the early mining and metallurgical texts used on the Pacific Coast. In 1876 Alfred Dewey bought rights to use a photo engraving process developed by Leo Bok of New York. Most of the line drawings appearing in the Press after that date were produced by that process. Dewey's proficiency in printing, as well as his interest in mining, brought business from the machinery companies, and his company developed into the principal supplier of promotional brochures distributed by the foundries.[43]

Dewey's pioneering of photo engraving in San Francisco, and his adaptation of new processes and machinery for production of his papers enhanced their look and value. As an editor he transformed the content of the *Mining and Scientific Press*. As previously mentioned, *The Scientific Press* of 1860 was not devoted exclusively to mining. Its columns contained "Household Hints," and all kinds advertisements, from hardware and patent medicine to drinking emporiums, dancing schools, and undertakers. This class of advertising soon disappeared, replaced by business listings of assayer supply companies, metallurgists, machine shops and foundries. Patent medicine gave way to ore crushers and "Household Hints" was supplanted by technical articles dealing with metal mining.

In this endeavor, Dewey was assisted by a young man being trained as a mining engineer, Charles G. Yale, son of Gregory Yale who wrote the first treatise on mining law published in the United States. Charles gave up his engineering career to help Dewey forge *The Press* into a trade journal serving the interest of American metal mining. They did that by recruiting the best writers of the day. Such men as Guido Küstel, author of *Concentration and Chlorination*, and Charles H. Aaron, whose books on assaying were published by Dewey and Company, frequently contributed articles. C. A. Stetefeldt, inventor of a furnace that bore his name, wrote about technicalities of ore reduction; and Almarin B. Paul, who had a mania for saving fine gold, wrote much about that subject. J. S. Phillips, author of *The Prospector's Companion*, was also a contributor; and noted metallurgist Melville Attwood, was a frequent visitor to the editorial room. Henry DeGroot, the first newspaperman to interview Mar-

A group of pioneer mining machinery men of San Francisco. Standing, left to right: Bob Kerr, E. A. Rix, R. B. Noyes, Samuel J. Hendy, fifth man unknown, A. J. Kittridge, John J. Hendy. Seated, left to right: first man unknown, Joshua Hendy, and William H. Reynolds.

shall, Wimmer, and others at Coloma after the discovery of gold in California, wrote extensively for the *Press.* Not only was he a pioneer miner, DeGroot was the best posted man on general mining affairs who could write. The first State mineralogist of California, Henry G. Hanks, wrote of the assay of gold, gold in black sand, borax, and California minerals. There were numerous other contributors, all leading miners and metallurgists. Little wonder, therefore, that equipment manufacturers packed the pages of the *Press* with advertising.

The first manufacturer to advertise in the *Mining and Scientific Press* was Lewis, Coffey & Risdon. They were little more than a boiler shop when they placed their ad on June 29, 1861. With superior service and products, and publicity from the *Press*, this company grew to be one of San Francisco's largest mining equipment firms, and eventually developed into the second largest shipbuilding yard on the West Coast. The Donahue Foundry, which became the Union Iron Works, placed its first ad on November 23, 1861. Others followed; the Pacific Iron Works, Fulton Iron Works, Aetna, the Miner's Foundry, the California Wire Works, and Joshua Hendy, just to mention a few. These companies proclaimed their excellence through the pages of the *Press*; "We make the best," "We lead where others follow," and "We guarantee the best."[44] Wild claims in some instances, the truth in most cases.

By the late 1860s, thanks to the Comstock, San Francisco had become the principal place in the United States for the manufacture of mining machinery, especially large plants. The Bay City had no rivals at that time, and its market was huge, the area comprising California, Nevada, Utah, Colorado, Montana, Arizona, Oregon, Washington, and Idaho. Canada, Mexico, Central and South America, Australia, New Zealand, Japan, China, and Russia were also added to this market in the 1870s. For nearly three decades mining men worldwide looked exclusively to San Francisco for their equipment, and they used the *Min-*

A graphic idea of the difficulties often encountered and as often ingeniously solved by miners all over Western America. Conveying an 800-pound cam shaft through dense spruce woods to a gold mine in the Bohemia District, Lane County, Oregon. (*Mining and Scientific Press*, August 7, 1897)

ing and Scientific Press to save them time and money in selecting the appropriate appliances. For that reason, foundries and machine shops considered the paper their special journal. The *Press*, of course, derived a large share of its revenue from their advertising. In response, the paper gratuitously plugged equipment manufacturers through "Foundry Notes," a week-by-week account of what was going on in the appliance trade, written by Charles Yale.

Covering the foundries and machine shops was a "rather unpleasant duty," according to Yale. "I had to visit these in rotation, interview the proprietor, and describe anything of interest in their shops. With a big piece of machinery I would often find part in the molding room, part in the machine shop, and part still in the draughting room. Taking the specifications and seeing such parts as I could, the descriptions were made."

Regardless, Foundry Notes served the *Press* well, giving it "first call" for line drawings of new equipment, illustrations that would bring its pages to life. Unpleasant as it may have been, by writing Foundry Notes Charles Yale publicized San Francisco's machinery manufacturers. Without his sloggings in the Iron District there would be little record of the city's impact on mining. Between 1871 and 1894 he left in prose and graphics an unmatched record of West Coast industry. We owe him a debt of gratitude, for most of this book is based on his work.

The *Mining and Scientific Press* presented an unmatched record of technical journalism. For thirty years Alfred T. Dewey directed the paper, but he was a man with divided interests, and the *Press* ran in the red most of the time. Then too, Dewey's sentiments, which passed through Ewers, often hurt the journal. The hydraulic mining controversy is a good case in point. Dewey's siding with state farm interests is reflected in the paper remaining strangely silent, much to the chagrin of the mining community. But he had able assistants who kept the journal on track: Almon

D. Hodges, Jr. (editor from 1869 to 1871), Charles G. Yale between 1871 and 1893, and H. F. Dexter. Throughout this period they focused on Western mining.

That changed in June 1893 when J. F. Halloran bought control of the *Press*. From then until December 1905, he and his editor, W. H. Storms, enlarged coverage to include the Rocky Mountain region. Most importantly, Halloran "put it on its feet financially." He made the *Press* valuable. The journal's coverage would be further enlarged when Halloran sold the paper in December 1905 to T. A. Rickard. Former Colorado State Geologist, Rickard was knowledgeable and had standing in the mining world. He was a prolific writer and an editorial genius, whose 1906 motto was, "Science has no enemy save the ignorant." An assistant stated he should have added "Mining has no friend like Science."[45]

T. A. Rickard controlled the paper for sixteen years, salvaging it following the 1906 San Francisco earthquake and fire. Although some of its advertisers never opened their doors following that disaster, the Press emerged a stronger publication, dominated by Eastern machinery dealers. Yes, by this time, San Francisco machinery dealers had dropped by the wayside, their demise attributable to a number of causes which we'll take up shortly.

By 1922 Rickard was looking for other editorial pastures, and on April 1 he merged the *Press* with its chief rival, *The Engineering and Mining Journal* of New York City, one of the first of many McGraw-Hill technical publications. For a while the combined periodical appeared under the title of *The Engineering and Mining Journal-Press*, but beginning with the July 1926 issue, the journal reverted to its original name. Nevertheless, the *Mining and Scientific Press* still marches on through the auspices of McGraw-Hill. Now let's look at factors contributing to the demise of San Francisco's machinery manufacturers.

The Comstock hit the skids in 1877 and its decline had an immediate affect on San Francisco foundries. Orders for stamp mills, pans, settlers, hoists and engines ceased. "Dull" was the word foundry managers used when asked about their business during winter of 1877–78. The slackness continued into spring, normally a time when manufacturers were gearing up for production. In fact, there were no heavy equipment orders during 1878; the foundries being kept alive by small contracts of "the miscellaneous kind," mostly for replacement parts. The following year was so slack that a number of smaller machinery dealers closed up.[46] Larger companies weathered the storm by producing equipment desired in other mining districts.

As the Comstock faltered, other regions in the West were coming to the forefront as mineral producers. Enormous deposits of argentiferous lead had been discovered at Eureka, Nevada, in Inyo County, California, in the Wasatch and Oquirrh mountains of Utah, and in Idaho. In 1875 refractory ore was discovered above the placers of California Gulch, and Leadville was born in Colorado.

Prospector-miners at first attempted to reduce these ores in Catalonian furnaces, six-foot tall refineries built of adobe masonry, fueled with charcoal made from oak, mesquite, or pine. Bellows provided the blast. Prone to melt-downs, these smelters could not keep up with production. Welsh, Cornish, and German smeltermen attempted to improve the situation by suggesting that furnaces be built of firebrick. That building material was durable but costly; not such a good idea in isolated portions of the Far West. San Francisco machinists came up with something better.

In 1868, at Darwin in Inyo County, California, they put an iron jacket around a furnace, and pumped water between the jacket and the crucible. The result was a cooler running furnace, increased production, longer smelter life, all that any smelterman could desire. San Francisco mechanics then took the idea a step further by designing efficient water-jacket furnaces that could be quickly transported, easily set up and maintained. The units could be used to smelt either lead-silver and copper ores. The era of the water-jacket furnace had opened; Arizona came alive as a copper producing region, and a stream of white metal poured forth from Utah, Nevada, and Colorado, glutting the silver and lead markets.

The water-jacket furnace pumped life into several West Coast machinery dealers. But that did not last, and by end of the 1880s their future again looked bleak. Their peril this time came not from depletion of ores, or closure of mining districts, but from competition for a market that at best was limited. Eastern manufacturers were moving into the mining appliance trade.

At first San Francisco companies were complacent about their competition. Boeyed by their isolation, they literally thought Eastern manufacturers did not understand and could not be made to understand the unique mineralogical conditions that had spawned the Western machinery trade. How short sighted. While Bay area foundries were gloating about their monopoly, David R. Fraser were quietly touring the Western mining districts, measuring and sketching quartz mills and ore reduction devices. The information he garnered gave birth to a small firm in Milwaukee known as Fraser & Chalmers, which subsequently became the world renowned Allis-Chalmers Company.[47]

Eastern manufacturers began penetrating the Western mining market in the early 1870s, not by setting up offices in San Francisco, but by moving their products through representatives such as Parke and Lacy. They gradually introduced their equipment over the course of two decades, then opened sales offices in California. In 1895 New York, Chicago, and Milwaukee companies had outlets in San Francisco, and were underbidding Bay area foundries for local work. When contracts for several large pumping plants went to Eastern competition without even an invitation to bid by local manufacturers, the foundries cried foul.[48] Certain types of engines, made in Eastern cities by concerns which make them a specialty, were sold on the West Coast cheaper than San Francisco foundries could make them. All this was made possible by the railroads.

Building of transcontinental railroads, at first heralded as a boon to Western foundry business, in the long run hurt Pacific Coast manufacturers. Completion of the Northern and Southern Pacific railroads made it easy for San Francisco foundries to ship equipment to Arizona, Utah, Montana, Idaho, Oregon, New Mexico, and several northern states of Mexico. On the other hand, these railroads made it just as easy to ship Eastern equipment westward, and by 1892 mines of the Pacific Northwest were being supplied from Chicago and Milwaukee. The railroad also opened Mexico to Eastern suppliers, forcing San Francisco foundrymen to drum for trade south of the border, trade which formerly came to them unsought.[49] The machinery market narrowed even more as transportation from eastern points became cheaper than from San Francisco.

Demontarization and the resultant depression of silver mining in the late 1880s adversely affected the foundries. Few, if any, new mines were opened, and many old ones stopped running. The situation was summed up by the *Mining and Scientific Press* of June 10, 1893: "Silver mines take more machinery than gold mines, for there are pans, settlers, etc., in addition to stamps, and with dullness in silver mining there is . . . dullness in the mining machinery trade."

Resurgence of gold mining improved the machinery market, but it wasn't anything like the days of the Comstock. Then too, gold mining equipment had to embody the most economical processes. Even in that trade San Francisco manufacturers had stiff competition, not only from Midwest manufacturers but from European companies which had control of the Australian and African markets. The English expended every effort to establish commercial relationships in Mexico, Central and South America, even going so far as to copy the designs of San Francisco foundries and put together stamp mills of all-iron construction, which were popular in tropical regions where wood frames quickly deteriorated.

In the best tradition of California salesmanship, the foundries endeavored to strengthen their Latin American ties by opening sales offices in Mexico and lobbying the Diaz regime. They exhibited their best equipment at the 1894 exposition in Santiago, Chile. The salemanship paid off. According to the *Mining and Scientific Press*, the president of one of the largest companies engaged in the manufacture of mining machinery reported "that had it not been for orders from Mexico he would have been obliged to close his works last July [1894]."[50] Nevertheless, the foundry business was so dull throughout the 1890s that the Union Iron Works and Risdon, San Francisco's largest equipment manufacturers, acquired land with deep-waterfronts, and went into shipbuilding. In the end they were gobbled up by Eastern steel interests. The remaining manufacturers either changed direction, going into the fabrication of compressed air appliances and hydro-electric components, or closed up. By 1910 the manufacture of mining equipment had shifted entirely to the Midwest and the East.

Now on to the foundries, their history, their products and services.

The Union Iron Works was founded in 1849 by Peter Donahue and his brothers, James and Michael. Other men soon joined the firm, bringing skills and drive to the company, and by 1866 it was the foremost foundry and machine shop on the West Coast. With the motto of "Quality rather than Quantity," the Union Iron Works excelled in the manufacture of steam engines and quartz machinery. The above engraving depicts company quarters on First Street, near the center of San Francisco, in 1866. Note geologist W. P. Blake's sign on the building. He served not only as an advisor to the Union Iron Works, but also had a sampling works on the property.

CHAPTER TWO

The Union Iron Works and the Pacific Foundry: Quartz Mills, Corliss Engines, Cornish Pumps, and Water-jacket Furnaces

SAN FRANCISCO'S first foundry was established by Peter Donahue and his brothers. Like many who came to California during the gold rush, they had only their skill and fortitude to build with. Their's is a typical rags-to-riches story. Michael, Peter, and James were born of Irish parents living in Glasgow, Scotland. Peter, the eldest and most prominent brother, was born January 11, 1822. Childhood for he and his brothers was one of hardship, apparently with little schooling, for Peter went to work in a Glasgow factory at age nine. That diragery was broken two years later when his parents emigrated to the United States.

The Donahues settled at Mateawan, New York, and Peter went back to work, first in a factory, then on a farm. In 1837 the family moved to Paterson, New Jersey, where the Donahue brothers appenticed themselves to the Union Iron Works. Peter learned the machinist trade and millcraft; James and Michael took up boilermaking and molding. In 1845 Peter returned to New York and found employment on the construction of a gunboat for the Peruvian government. He accompanied the completed vessel to Peru as an assistant engineer. It was while in South America that Peter learned of the California gold rush.

Peter Donahue left Peru in 1849 to seek the "Elephant." It was a voyage he almost didn't survive. During a layover in Panama he was stricken with Isthmus fever, probably malaria. After convalescing several months at Tobago, Peter boarded the steamship *Oregon*, bound for San Francisco. That voyage too was an adventure. When the vessel's engine broke down, Peter Donahue repaired it. For that service, the grateful owners rewarded Peter with $1000, and urged him to stay with the company. He declined; the lure of gold drew him on.

Landing in San Francisco, Peter immediately headed for the diggings. After six months of unsuccessful placering, it dawned on him that perhaps fortune lay with what he knew best. Donahue returned to San Francisco, and joined by his brothers, erected the town's first foundry in the sand hills near the beach, a spot destined to become the intersection of Montgomery and Jackson streets. With a crude furnace, the blast for which was produced by two blacksmith bellows, they poured the first iron casting turned out on the Pacific Coast, "a steady or spring bearing" for the ocean-going steamer *McKim*. Fifty cents a pound was their charge.

The Donahue foundry may have been primitive, but there was an immediate demand for its goods and services, and the business grew rapidly. From time-to-time the establishment was enlarged and improved, various departments added. In 1856 James and Michael sold their interest in the property to brother Peter, who immediately built of brick a spacious 50,000

Irish machinist Peter Donahue died a wealthy man on November 26, 1885. Besides establishing San Francisco's first foundry, he pioneered public utilities in California, and was active in railroading. He was twice married; the first time in 1852 to Mary Maguire. She did not live long, and in 1864 he married Annie Downey, sister of California Governor J. G. Downey.

square foot machine shop and foundry in what was called Happy Valley, between Mission, First and Fremont streets. There, Peter carried on the business in the family name, the Donahue Foundry.

The mechanical branch of Donahue's foundry did not do well. Nobody really knows why it faltered. Some assert it was due to incompetent agents, others say it was because of lack of personal attention. There may be some validity to that statement, for Peter Donahue was a civic minded individual, engaged in city and state enterprises. In 1852 he organized the first gas company for street lighting in San Francisco, and in 1860 he was appointed treasurer of the San Francisco & San Jose Railroad. The following year he founded the Omnibus Street Railway, San Francisco's first street car line. In 1862 he became one of the charter members of the Union Pacific Railroad. No doubt about it, Peter Donahue was busy, probably obsessively so.[1]

To ease the burden of managing the foundry, Donahue formed a partnership in 1863 with H. J. Booth, owner of the Marysville Foundry, and C. S. Higgins, formerly of Freeman's Express. A new corporation was formed, known as Donahue, Booth & Company. It did business as the Union Iron Works, a name adopted in honor of where Donahue learned his trade.

Booth brought to the firm enterprise, talent and integrity. An excellent financier and a thorough mechanic, he successfully carried on the foundry business for about twelve years. C. S. Higgins was also well known as a "gentleman of active business qualifications." But jealousy as to supposed rights, and real or imaginary conflicting interests, prevented a harmonious relationship between Booth and Higgins, and the firm suffered because of it.

With each passing year Donahue became more engrossed in railroading. Selling his interest in the Union Iron Works to Booth in 1865, Peter embarked upon a program of wheeling and dealing in transportation. By this time he and a partner had acquired a two-thirds interest in the San Francisco & San Jose Railroad, which they sold in 1870 to the Central Pacific for $3,250,000. That year Donahue acquired a controlling interest in the San Francisco & Humboldt Railroad, out of which he created the San Francisco & North Pacific Railroad, its branches and extensions lying in Marin, Sonoma, and Mendocino counties. In the years before his death on November 26, 1885, Peter Donahue accumulated great wealth. Now let's backtrack to 1865 to pick up the development of the Union Iron Works.

Upon Donahue's retirement, Booth immediately brought in G. W. Prescott, who had managed the Marysville Foundry since 1859. He in turn bought out Higgins. Another man now entered the picture, Irving M. Scott, Donahue's superintendent since 1863. A young man of slight financial means, Scott purchased a small share of the Union Iron Works. The new firm of Booth, Prescott, and Scott was formed on March 16, 1865. Although Booth had sound financial sense, and Prescott was a knowledgeable mechanic and foundryman, Scott brought drive and inventive genius to the company. It was a partnership marked for success.

Scott was born December 25, 1837, on the farm of his Quaker parents, John and Elizabeth (Leittig), at Hebron Mills, Baltimore County, Maryland. Irving grew up working on the family farm where he developed an early interest in the mechanisms associated with grist mills. His upbringing may have been rural but his education was urban. Scott attended Baltimore public schools, and upon graduation studied drafting at the Baltimore Mechanics' Institute. That skill landed him a position in 1854 with Obed Hussey, inventor of the reaping machine.

Although Scott learned wood and iron working from Hussey, he longed for a career in mechanics. His opportunity came in 1857, when he obtained a position with Murray & Hazelhurst, a large Baltimore iron works. There, Scott assumed the duties of draftsman in the machinery department. So skilled did Scott become that his employer transferred him to design, and placed him in charge of the construction of stationary and fire engines.

In 1860 Murray & Hazelhurst assigned Scott the duty of delivering a steam engine to Peter Donahue in San Francisco. It was one of those rare opportunities that changes one's life. Scott didn't much care for the rough voyage around Cape Horn, but he was enthralled with California's robust vigor, captivated by San Francisco's vitality. Here was a country he could "grow up with." At conclusion of his project, Scott stayed in California, entering the employment of Peter Donahue as a draftsman. In 1862 Scott transferred to the Miner's Iron Works, taking charge of its drawing department. He was there only a short time, returning the following year to the newly organized Union Iron Works, where he was made superintendent.

A man of exceptional enterprise and public spirit, Scott had a rare knack of instilling his ideas in others. Well informed on topics not connected with his business, Scott was a thinker who expressed his ideas clearly and concisely. He was a fine public speaker, a writer of more than ordinary ability, and he was fond of art and literature; a man of refinement and education. Irving Scott was a Renaissance Man in the strictest sense of the word. No individual was more prominently identified with the early industrial interests of the Pacific Coast than Irving M. Scott.[2]

Scott's ability as a mechanical engineer was quickly recognized, and he designed all sorts of machinery for general and special purposes. He excelled at applying mechanical arts to the emerging field of mining.[3] With Booth attending to the contracting and financial departments, Scott supervised the mechanical end of the foundry. It was a successful combination, and in a short time the Union Iron Works was operating day and night. As the volume of work increased, larger machinery than had hithertofore existed on the Pacific Coast was installed. Within several years the company had nearly 300 employees at work on twenty-three lathes, eight planers, twelve drills, several gear cutters, boring mills and slotters. A fifteen ton steam hammer, sixteen forages, two cupolas of eight ton per hour capacity each, were added. Still the work accumulated to such an extent that the company was compelled to enlarge its operation.

Irving M. Scott went from draftsman to general manager, and then on to president of the Union Iron Works. Besides helping to revolutionize mining equipment, he foresaw a huge maritime commerce between San Francisco and the rest of the world. Between 1883 and 1900 he built the company into the West Coast's premier shipyard. He died on April 28, 1903, leaving a widow, Laura Harde of Covington, Kentucky, and two children.

The company had capabilities to produce the largest castings for every description of machinery. In 1865–66 it built for the San Jose Railroad Company, the *California*, the first full-size locomotive constructed on the Pacific Coast. Others followed in rapid succession: another one for the same company; one of the same weight (fifty tons) for the Sacramento Valley Railroad, and one for the Central Pacific Railroad; and a locomotive of eighteen tons for the Pittsburg Railroad, together with eight cars of three tons each.

Because of its maritime and locomotive work, the Union Iron Works dominated West Coast production of portable and stationary steam engines. Of the latter, the company produced two industrial workhorses:

Although the *California* was the first full-size railroad engine built in California, it was not the first locomotive fabricated in the state. The Vulcan Foundry turned out a small locomotive for an Oregon line in 1862. Built for the San Jose Railroad, the *California* made its trial run on August 30, 1865, drawing a large number of cars a portion of the distance at the rate of a mile a minute. The company followed by fabricating locomotives for a number of California railroads, including engines for the Pittsburg Road, which hauled coal from the Mt. Diablo coal mines to steamer landings on the San Joaquin River. The engines used on the latter road were designed to overcome a maximum grade of 276 feet per mile. (*Mining and Scientific Press*, October 16, 1869)

the Corliss engine, an American favorite; and the British giant, the Cornish engine. Engine accessories, such as cut-off valves, became big business.

In 1867 Scott and his chief draftsman, William R. Eckart, patented a slide valve governor, a simple contrivance for controlling the speed of steam engines, whether working under a load or not. The ingress of steam was connected directly with the governor, making it supreme, shutting off or letting on steam as circumstances required.

It was, however, the design and manufacture of quartz mills and amalgamation works that was the pride of the Union Iron Works.[4] In that branch of trade Irving M. Scott's hand was on every item. No better example can be found than his contribution to the refinement of the California stamp mill.[5] But first a little more background on stamp mill development.

As soon as stamps began to fall on California gold quartz, men went to work to improve the crude crushing devices. Unfortunately, the first innovations are difficult to trace, for early stamp mill practices were marked with little systematic research, much crudeness of construction, and secrecy of methods. Previous to adoption of the round revolving stamp, numerous departures were made from the original imported type. One stamp mill style came into vogue around Marysville, California, having square stamps fastened to stems by means of a taper driven into the stamps' heads. The rectangular tappet was held by cross keys, one for each stem; the stems cross-slotted to engage the keys. Some of these stamps weighed as much as 1,000 pounds each. The shoes were held by a square tapered shank.

The *Mining and Scientific Press,* of October 21, 1893, claimed that C. P. Stanford invented the round stamp, but it was Isaac Fisk, an engineer in Stanford's employ, who made the stamp revolve. He did not patent the idea, however. Joseph L. Laird tried to. On July 4, 1854, he applied for patent for an improvement in ore stamps. "The arrangement of the lifters by being so placed as to operate on the periphery of the tappet head for the purpose of giving the stamps a partial rotation without requiring other mechanism in the manner set forth." Laird's patent was found in-

No where in the United States did steam-power play such an important role in general industry as on the Pacific Coast. The high price of labor compelled its use wherever possible, in mining, drainage, agriculture, manufacturing, and transportation. The Scott & O'Neill compound engine, shown above, was developed to drive the machinery of the Union Iron Works. First built about 1880, these engines found wide application in mine haulage, and they were later adopted to power cable rope ways and railroads. (*Mining and Scientific Press*, January 18, 1890)

valid for "lack of novelty," and thousands of stamps operated with this improvement, free of royalty.

After the round stamp was introduced, Zenas Wheeler devised a means of holding the tappet by screwing it upon the stem. Later, when having a large stamp mill built at the Miner's Foundry of San Francisco, he accepted the advice of proprietor H. B. Angell, and had it made with a gib and two cross keys in the tappet. About 1865 or '66, Irving M. Scott casted the gib into the tappet, making it shorter and entirely enclosing it at the ends, thereby securing it against maladjustment, as well as simplifying its construction. Thus Scott set the stage for the introduction of what is known as the "high mortar," an important modification to the California stamp mill.[6]

The high mortar was the result of Scott's careful collection of data pertaining to gold milling. According to the *Mining and Scientific Press* of September 27, 1902, shortly after assuming superintendency of the Union Iron Works in 1863, Scott "visited all points of interest" on the Mother Lode, and examined, measured and sketched various mortars in use, and the kind of gold ore being milled. He found the mortars were no more than wooden troughs lined with sheet iron, five to seven inches deep, in which dies were set. Mortar height always varied according to the ore being worked.

Low mortars were used for quartz having coarse gold, high mortars for ore with extremely fine gold, and where gold varied — "some flakes and some grains" — mortars of medium height were utilized. Apparently there was some sort of crude law governing the kind of mortar to be used. Millwrights who fabricated the mortars, however, could not tell Scott the reasoning behind the various mortar heights; they had discovered the true height and shape merely by experimentation.

Scott returned to San Francisco to ponder the different methods of crushing quartz. He correlated his sketches and measurements with mortar output, and upon the drawingboards in the office of the Union Iron Works, he designed various mortars for gold working, for dry crushing silver, for wet crushing and for amalgamating in the mortar, and for amalgamating when gold was carried in silver ore. Scott also experimented with mortar sizes. He ran exhaustive tests on mortar prototypes, and found that mortars used to amalgamate had to be wider than those used to simply crush ore, or the dropping stamp would cut the amalgam off the back of the mortars, or off the lip of the copper plates. Narrow mortars used in wet crushing restricted the tonnage of ore pushed through the stamps. Mortars were not the only component of stamp mills that Irving Scott altered.

The rise and fall of the mill's stamps were timed by a shaft bearing a series of cams. Because the shape of these cams affected rapidity of stamp fall, it was not long before millmen began experimenting with their configuration. A Dr. Southgate invented a single-arm cam with a hub on one side. He claimed he could get more drops per minute with a stamp, with less jar or blow, by the single-arm cam, hub on one side and stem close to the shaft, than by any other device. There was one thing wrong with Southgate's cam. At high revolution the cam and tappet developed serious wear, and it was difficult to keep the cam tight on the shaft. This was remedied by Irving Scott.

It occurred to Scott that "there was no reason why the cam should not be built with double arms and the hub placed on one side, getting the stem just as close to the side as possible, and giving a blow very much less on the tappet." The Union Iron Works design department carefully worked out the idea, and put double arm cams on a mill at the Rappahoe Mine. Except for some wear, the mill worked with little vibration. Scott then tackled the wear problem.

He first attempted to remedy the surface wear of tappet and cam by chilling both cam and tappet surfaces. That didn't work, as neither could be satisfactorily adjusted after they were chilled, and grinding changed their angles somewhat. Finally, steel faces and rings were put on the tappets, and steel sheaths sweated over the edge of the cams where lifting was done. Eventually both methods were discarded in favor of a double-arm cam with a hub on one side, which overcame all difficulties. According to Scott, "if made of proper iron, this cam lasted a sufficient time for all practical purposes."

Scott contributed more to the mechanics of metallurgy than any other man on the coast, improving the mechanical details of nearly all pans and settlers, making suggestions of value to inventors while their machines were being fabricated. His research pushed the Union Iron Works to the forefront in design and manufacture of quartz mills. When the Union Iron Works began manufacturing stamp mills, the crude devices crushed only half a ton of quartz dry, and one and a quarter tons wet, per day. By 1875 Scott had mills handling two tons dry, and up to five tons wet.[7] He did not lag behind in producing amalgamation works. In conjunction with the Golden State Foundry, he fabricated Wheeler and Randall Tractory grinders and amalgamators, as well as Prater concentrators.[8]

Of all amalgamation equipment produced by Union, the Wheeler pan was the most successful. No quartz mill in California or Nevada was considered complete without them, and they were introduced to Australia by San Franciscan Robert Farmer, who employed Wheeler pans to reduce tailings in the Bendigo mining district. The pans operated so well that he organized a firm in Sydney to sell them to Australian mining companies.[9] Thus, the Union Iron Works gained a foothold in the foreign mining market.

Mining companies sought Booth and Scott's expertise at building hoisting and pumping engines. Eight-inch cylinder engines suitable for hoisting were produced, as were Cornish pumps with twelve-inch cylinders and twenty-four-inch strokes for deep pumping on the Comstock and in Idaho. They also built large Corliss engines capable of driving pumps and mills of forty stamps or more.

It did not take long for foreign companies to recognize that San Francisco-made equipment was superior to any built in England or the eastern United States. By 1870 the firm was turning out stamp mills and amalgamation works for mining companies in Mexico, Central and South America; and in spring 1873 Union shipped to Japan a complete quartz mill with pans and four Hendy concentrators. Foreign preference for Scott's appliances was due in large measure to his policy of incorporating the latest improvements into Union's equipment.[10]

H. A. Ward, superintendent of the English-owned Midas Mining Company near Virginia City, Montana, after "thoroughly acquainting himself with both Eastern and California-made mining machinery," insisted on buying Union equipment, notwithstanding the extra expense of shipping thirty tons of equipment from San Francisco (30 cents per pound compared to 12½ from St. Louis). Ward's choice was a 15-stamp mill, including a forty-five-horsepower engine, a Blake crusher, two Wheeler & Randall pans, three Attwood agitators, three Hendy concentrators, a No. 3 Excelsior pump, smelting furnace, and retorts. Another 6-stamp mill and miscellaneous equipment was also ordered by the same company.

All engines shipped to the Midas Mining Company were manufactured to order in San Francisco by the

Figure 1.

Figure 2.

Wheeler & Randall's tractory grinders and amalgamators were included in the class of devices known as Washoe pans. Their prototype was actually the British tractory flouring mill which produced more flour than a four-foot flat mill-stone. Redesigned for amalgamation, and tested at the Bacon Mill in Silver City, Nevada, they were mass produced by the Union Iron Works and the Golden State Foundry. Figure 1 shows the pan with the base of the grinding surface (conoid) upward, which allowed the quicksilver to remain nearer the center of the pan.

Figure 2 represents the Wheeler pan with base of the grinding surface downward, which allowed the concentration of quicksilver at the circumference of the pan. These pans utilized open mullers to keep the pulp in contact with the grinding surfaces. The Wheeler and Randall pans were eventually manufactured by all San Francisco foundries.

Union Iron Works.[11] They were Scott and Eckart cut-off engines, 18 x 42- and 20 x 42-inch models, proclaimed for their beauty and quality of workmanship. "Not only are they solidly and finely built," observed the *Mining and Scientific Press*, "they are also very tastefully and appropriately decorated." On one side of the engine governor, Union foundrymen placed an American eagle and shield. The British lion and shield was placed on the other side of the governor. On the engines' sides were fixed the arms of England and the United States, and combined motifs were placed at the ends of the cylinders. Crossed flags of both nations decorated the bed plates.[12]

For ten years the firm of Booth, Prescott, & Scott prospered until it was universally acclaimed the best equipped foundry and machine shop on the Pacific Coast. On June 8, 1875, H. J. Booth retired and the company became merely Prescott, Scott & Company. The new company, however, included H. T. Scott, Irving's younger brother, who had acted as purchasing agent for eight years. Upon Booth's retirement, he took a more active role in company finances, allowing Prescott, who was now senior member of the firm, time to travel worldwide gathering technological information and promoting the company. Irving Scott continued supervising the mechanical end of the business, distinguished himself by improving the mechanical details of all varieties of quartz machinery.

While the Union Iron Works was the largest foundry in San Francisco by 1875, there was nothing pretentious about the establishment. Its buildings were functional, housing eleven departments, each under the charge of an experienced foreman. The drafting department employed eight draftsmen under the charge of Eugene O'Neill, who had worked his way up from the machine shop. The pattern shop, the next department in the long production line, consisted of a battery of planers, lathes, band and circular saws, and miscellaneous machinery run by eighteen men supervised by William McConnell. Their duty was to speedily turn out patterns for prototype development, which upon approval were sent to the foundry.

Connected with both the pattern shop and the foundry was the flask shop, in charge of Joseph Burke, where all wooden flasks and wood work pertaining to a large foundry was promptly and expertly prepared.

The most extensive branch of the business was the foundry, consisting of 127 molders under supervision of Henry Dimmick. That department was fitted with two large core ovens, three cupolas with a combined melting capacity of 64,000 pounds of iron in two and a half hours; and there were pits and appurtenances for casting the largest cylinders. All products of this department were sent on to the machine shop.

The machine shop was designed to handle the requirements of large equipment being manufactured for the deep mines of the Comstock Lode. There were thirty-four lathes, from twelve to twenty feet, thirteen planers from six feet wide by twenty-four feet long to ten feet wide by thirty feet long. Both lathes and planers were the largest on the Pacific Coast. Six slotters; fourteen drills from ten to ninety-six feet; two vertical mills from four- to eight-feet; four horizontal boring mills, three bolt cutters, two nut tappers, two nut facers, six shapers, four pulley lathes, two spur gear cutters from two and a half inches to twenty-six feet; two bevel gear cutters from two and a half inches to ten feet; also a complete assortment of milling machinery, small tools, cranes capable of handling thirty tons, cars, and car tracks. The department was supervised by William Worth, an old hand.

The blacksmith shop, under control of Patrick Hayes, consisted of forty men working at seventeen forges and two steam hammers. Supplied with cranes and small tools, they turned out all kinds of forgings.

The boiler shop was the newest addition to the company, being erected in 1874. Boiler construction was done under a new system of punching, accurate enough to eliminate the necessity of pin drifting, which was considered injurious to boilers. There, all

This engraving from the Union Iron Works Catalog No. 5 (1898) depicts a modern 10-stamp mill. The battery is of the design known as the back-knee type, driven by belt and tightener from a shaft located upon the battery frame sills below the mortars. The drawing illustrates not only the battery and its various parts, but also shows the ore-bin gate, feeders, copper apron plates and water piping, all in their relative positions.

sheet iron was planed, rolled and caulked. Seventy-four men, supervised by Levi F. Cole who had been with the company since 1864, turned out superior work.

The tool shop was responsible for the storage and repair of all tools used by company personnel. Tools were released to workmen, who assumed responsibility for their condition. Mechanics returned their tools at the end of a shift or upon completion of specific jobs. In that way everything was tracked, inventory kept up, and tools repaired. Frank A. Davis was in charge.

A screen department turned out fifty different varieties of screens for quartz mills. They were made of Russian iron, galvanized iron or zinc. Holes were punched in these plates from half an inch diameter down to that of a number ten cambric needle. The department's special machinery was designed and constructed at the works. A Mr. Buxton was in charge.

A shipping department under charge of Robert Peckham, weighed, crated, and shipped the company's manufactured products. All orders were carefully tracked so that injured or lost pieces could be repaired or replaced.

Like large corporations of today, the Union Iron Works had a personnel section, then called the labor department, which hired unskilled labor to assist mechanics and foundrymen. A police system looked after security and maintained order among the company's 476 employees.

Last but not least, the Union Iron Works had an apprenticeship system designed to provide a pool of trained mechanics. The firm took in boys at age seventeen and trained them for four years, paying them $4 per week for the first year; $6 per week the second year; $8 per week the third year; and $10 the fourth year. During this probationary period proprietors reserved the right to discharge any boy for incompetency or lack of punctuality. If an apprentice left the establishment without permission during that time, they forfeited all pay due them up to the time of departure. Most boys proved themselves attentive and industrious, for no shop would hire an apprentice from another shop unless he presented a written discharge. This established trade rule prevented boys from roaming from one shop to another. By the expiration of apprenticeship, management had a good estimate of a youth's worth, and if they had proved good workers they were kept on as journeymen. The apprentice system rewarded ability and good behavior with promotion. In 1874 the Union Iron Works employed 340 men, of whom seventy-two were apprentices.[13]

Next to quartz mills and amalgamation works, Union excelled in the construction of pumps. Touted as "a California invention of superior merit," its Hooker Excelsior pump, put on the market in 1866, had wide marine applications. As far as mining went, it was limited, serviceable only to depths of 400 feet.[14] Deep mining required Cornish pumping engines in several variations, and the Union Iron Works made these engines in both single and double-acting models, and in over- and under-beam units, the latter being a specialty.[15] Because there were only three firms in San Francisco equipped to manufacture or alter these ponderous hydraulic systems (Pacific Iron Works and Risdon being the others), Union received sizeable contracts for Cornish engines from Comstock mines. In summer of 1875 it constructed for the C & C shaft the first direct acting pump ever built on the West Coast.

This Cornish pump was a handsome machine, simple in design. The cylinder was forty inches in diameter with an eight-foot stroke. The steel piston rod was twenty-two feet long, eight inches in diameter. The cylinder was surrounded with a jacket to prevent condensation of steam, and had every appliance to economize fuel. Steam and exhaust valves were brass, fitted with steel rods.[16]

Union also did subcontracting for other foundries engaged in building machinery for the Comstock. In February 1876 it casted for the Pacific Foundry the largest cylinder ever made on the Pacific Coast, a part of a 2,000 horsepower Cornish engine being installed at the joint shaft of the Chollar, Hale & Norcross and Savage mines. This pump had the capacity to lift a column of water from a depth of 4,000 feet. Total weight of the equipment was 700 tons.[17]

Silver production on the Comstock Lode began to wane in 1876, and so did contracts for heavy machinery. What remained to keep San Francisco's foundries going were numerous small orders from mining districts developing over a large geographical area. From the Cariboo mines of British Columbia, Union received an order in 1878 for a 20-stamp mill. Its largest contract that year was for an 80-stamp mill for the Homestake Mine in the Black Hills — George Hearst was not about to let his beloved San Francisco down.[18]

Regardless of slackness in the equipment trade, Union continued to upgrade its machinery and develop new lines. In February 1880 it introduced to the mining world the Dodge Crusher and Pulverizer. Invented by Miles B. Dodge, and tested for some time in Colorado, this rock breaker was advertised as being capable of doing the work of a 10-stamp mill.[19]

Mining slumped in the West with the ending of the Comstock boom, and by the early 1880s mineral extraction technology had wrought a pronounced

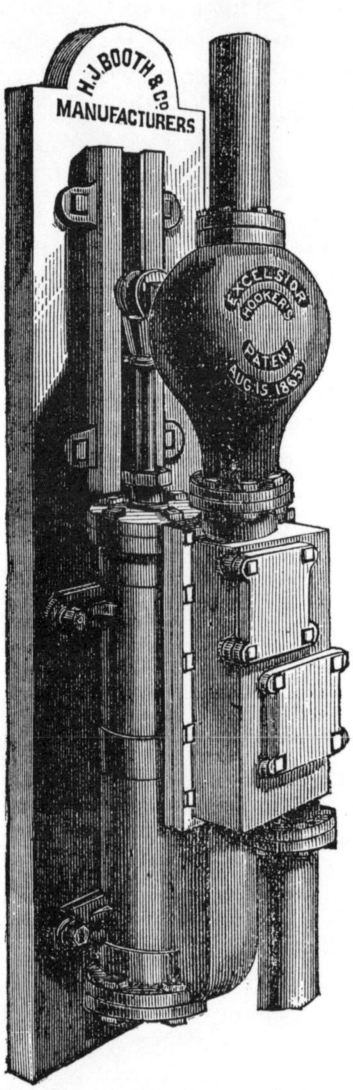

The Excelsior Challenge Double-Acting Force Pump, known as the Hooker Pump, was a California invention of superior merit. Produced by the Union Iron Works, the pump was substantially made, and adaptable to any power. The pump was warranted by its manufacturers to raise more water than any other of equal size, driven by the same power. It could be readily placed in position, was of simple construction, so much so that it was said to be almost impossible to malfunction. The pump could also be run at high speed, a quality which made it attractive for fire engine application. The power model of the pump, shown above, was adaptable to mine work at depths of 400 feet. A hand actuated brass model was available for ships and steamers. The Hooker Pump was selected by the California Steam Navigation Company for use on the *Capital*, a steamer put on to the Colorado River in 1866-67. (*Mining and Scientific Press*, June 16, 1866)

What the Union Iron Works excelled at was the fabication of pumping plants capable of giving a duty of 70,000 to 80,000 gallons by the Cornish rule, an example of which is shown on the opposite page.

change in equipment. Appliances were becoming compact, light and easily transported; new processes for extracting precious metal were at hand. The machinery market was decaying. To make matters worse, Eastern companies were penetrating a market long dominated by San Francisco firms. Many city foundries fought the competition by opening offices in the East and improving their service. They even lobbyed in Mexico, Central and South America. This of course bought time, but in the end it was a losing fight. Irving Scott saw little sense in combat and prepared to change the course of his company.

In 1880 Scott embarked upon a cruise which took him to the world's major shipyards. He returned to San Francisco convinced that ocean transportation would one day outpace the development of railroads. Scott clearly saw that the future of the Union Iron Works lay not in mining equipment but in shipbuilding. A new corporation was formed with George W. Prescott as president, Henry T. Scott vice president and treasurer, and Irving Scott as general manager. Scott gave direction to the enterprise by bringing in George, James, and John Dickie, three brothers from Scotland with a long history of shipbuilding. John and James had established the Dickie Brothers shipyard in San Francisco in the 1870s, and George had charge of design and construction of marine engines at the Risdon Iron Works. Like Scott, they saw great potential in West Coast shipbuilding.

Securing additional capital in 1883, the Union Iron Works acquired twenty-five acres of land at the Potrero near the Pacific Rolling Mills. There, they built an extensive complex, including a shipyard and drydock. A new boiler making shop, 200 x 90 feet, was built, along with a smith, 40 x 200 feet, a pattern shop 50 x 150, and a foundry 90 x 200 feet, and a copper and brass foundry, and a 40 x 130 foot machine shop. The installation included a number of traveling cranes. Now the largest plant on the Pacific Coast, with as much capacity as all the foundries in San Francisco combined, Union Iron Works embarked upon an ambitious program of constructing iron and steel ocean-going vessels.[20]

One of the first jobs of the new yard was the fabrication of the first all steel merchant ship ever built in the United States, the collier *Arago*. In summer of 1886 Union launched the *Balboa*, a steel water boat for the Panama Railway Company.[21] The real prize, however, came when the U.S. Navy awarded Union the contract to build the cruiser *Charleston*. The first United States warship built on the West Coast, this cruiser was launched on July 19, 1888.[22]

Union Iron Works constructed large numbers of beam engines of various sizes and kinds. The above beam engine, with Corliss valve gear, was built for the Starr Flour Mills near Port Costa. The engines were of 3,000 horsepower, and was the heaviest in the world applied to grinding flour. Similar engines were utilized to power steamboat, and for mine drainage. The Union Iron Works carried patterns and drawings for all kinds of beam engines from 100 to 2,000 horsepower, thus making the company one of the prime sources on the West Coast for hydraulic equipment. (*Mining and Scientific Press*, August 16, 1890)

By spring 1887 Union's work force had grown to between 800 and 900 men, 200 of which were employed at the shipyard. A keel of a second cruiser, the *Baltimore*, was laid in October 1887,[23] and launched October 6, 1888. Construction of a third cruiser, the *San Francisco*, got underway in December 1888, and was launched in late October 1889. The *Monterey*, the fourth cruiser, was launched in early May 1891, and a fifth cruiser, the *Olympia*, was launched November 5, 1892.[24]

It was not all shipbuilding at Union, however. In 1886 the firm was awarded a contract to build the great steel dome for the Lick Observatory on Mount Hamilton. About this time, Union launched a campaign to manufacture and market the Ide Engine, a steam engine that was awarded first prize for engineering by the Franklin Institute in fall of 1885. This compact, simply constructed, powerful engine was guaranteed not to vary its speed more than one percent no matter what load its was working under.[25]

Union continued turning out mining equipment, principally stamp mills. In spring of 1887 it built a 60-stamp mill for a company in Butte, Montana, and that summer, a plant of 80-stamps with thirty-two concentrators (sixteen Frues and sixteen Triumphs), with the necessary rock breakers — in all 400 tons of machinery — was completed for the Union Mill and Mining Company of Alaska. While most mining equipment was diminishing in size, the capacity of Union stamp mills had grown.

But the Union Iron Works continued to focus on shipbuilding. On October 26, 1893, it launched the battleship *Oregon*, the largest vessel in the U.S. Navy.[26] Three years later, Union attained another first in shipbuilding when the Japanese government awarded the company a contract to build a second-class, high speed cruiser for its Imperial Navy.[27] Christened the *Chitose*, this vessel made its trial runs in the Santa Barbara channel in January 1899.[28] But we have gotten ahead of our story.

By 1890 there was movement at Union to split its mining equipment division from shipbuilding. That became reality in late January 1892, when it purchased the ailing Pacific Iron Works, San Francisco's

The above differential compound, direct-acting beam engine was produced by the Union works for pumping purposes. Valve motions were controlled by Davey differential gearing, so that the engine automatically adapts itself to the resistence or amount of water, and the speed or number of strokes. The valves were of the balanced poppet class. Condenser, air and feed pumps were placed in a pit formed of supporting masonry. To the left is shown the top section of a common pump rod and the mouth of a pump shaft, indicating the manner of making connection.

second largest foundry. At this point, we need to step back in time and look at the growth of this manufacturing company.

Like the Union Iron Works, the Pacific Foundry and Machine Shop was a pioneering company. Founded in 1850 by Robert Egery and David B. Hinckley, it was located in a fifty by sixty foot, one story, frame building at 127 and 131 First Street, between Mission and Howard, in what would become San Francisco's foundry district. Three years later the interests of the founding partners were purchased by New Englander E. B. Goddard, who erected an additional three story building, thirty by sixty feet. The structure was enlarged in December 1858 to house a casting department.[29]

The Pacific Foundry and Machine Shop started out doing general foundry work and maritime repair. That changed in 1862 when Ira P. Rankin, a Scot boilermaker, and A. P. Brayton, a molder from Rhode Island, bought into the business. Two years later, on April 25, 1864, the company was incorporated under the name of Pacific Iron Works, with a capitalization of $200,000, divided in 1,000 shares at $200 each. The Board of Trustees was composed of W. G. Heslep and John Stone (of Tuolumne County), and San Franciscans J. M. Wooster, B. D. T. Clough, and W. MacDonald.[30]

There were many forms of engines working on the Cornish pumping system, the single and double-geared engines being the most common. Direct-acting engines were made in a variety of ways, and connected directly to the main bob which actuated the line of pump rods. The engraving at right shows a direct-acting compound pumping engine with O'Neill rotative valve-motion and cut-off, a style of under-beam engine made by the Union Iron Works. In this design, the cylinders were placed side-by-side, on a heavy bed-plate to which were bolted the guide-plates for the wrought-iron cross-heads. The steam-chests were cast separate, and bolted to the cylinders. Connecting rods of each engine were fastened directly to one end of the beam. This particular engine is fitted with jet condenser, and independent bucket plunger air-pump. The engine was made in several sizes. (*Mining and Scientific Press*, December 4, 1886)

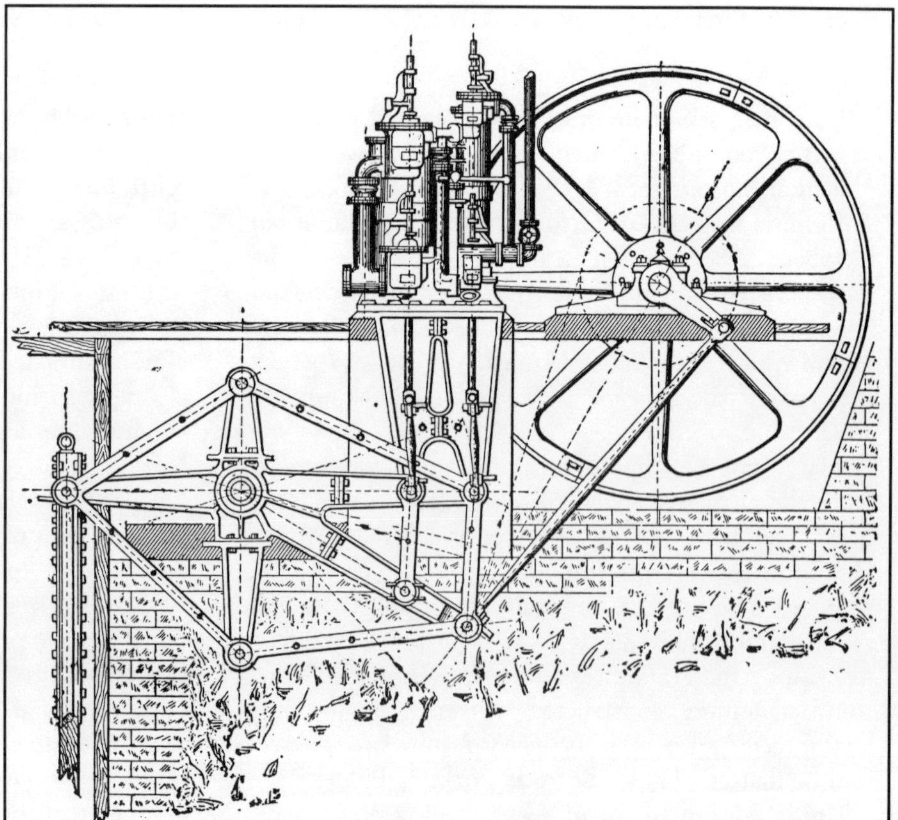

The Ball steam stamp, at right, was patented by E. P. Ball of Chicopee, Massachusetts, and manufactured by the Holyoke Machine Company of Holyoke, Massachusetts. First introduced into the gold mines of North Carolina, these mills worked successfully until the mines were exhausted. About 1857 the Ball Mill was put to work processing copper ore at the Copper Falls Mining Company of Michigan. Many improvements were made to the Ball Mill over the years, increasing their efficiency and durability. The mill shown is the smallest of the 1877 version, having a daily capacity of between fifteen and twenty tons of ore, when using the finest screens. A larger version, with six heads, was built for the Calumet and Hecla Company. It was capable of crushing 800 tons every twenty-four hours.

While there was no attempt on the part of the manufacturers of the Ball Mill to introduce it on the Pacific Coast, the principle behind the mill was utilized by the Pacific Iron Works in the manufacture of the Wilson Steam Stamp Mill, shown on the next page. (*Mining and Scientific Press*, June 2, 1877)

Focusing on repair of marine engines and ship hardware, the firm built up an impressive clientele which included the Ben Holliday Steamship Line and the Pacific Mail Steamship Line. This repair work soon blossomed into full scale fitting out of ships, and by 1865 Pacific employed 150 men who kept three cupolas going, melting about twenty tons of iron at any one time.[31] That year the company built machinery for the Steamer *Del Norte*, and overhauled the *J. T. Wright* which cruised the north Pacific in service of the Russian Telegraph Company.

With opening of trade with China and Japan, and the increasing coastal business which created a demand for new vessels, the firm branched into design and construction of ocean-going vessels. Concurrent with its maritime business, the Pacific Iron Works turned out custom work for mining companies, building engines for the Comstock. In 1860 Pacific introduced the Bryan Quartz Mill of five batteries, and followed shortly thereafter with the Varney amalgamation pan.[32] Both appliances were widely accepted, particularly in Kern County, California, and in Nevada. The Dall Mill in Washoe Valley used Pacific equipment exclusively.[33]

In spring of 1867 Pacific made a novel departure in the engineering of quartz mills, when it built under the patent of James and Nathan Condict, Jr., a mill where steam was applied directly to the stamps. This was not a new idea, for E. P. Ball of Chicopee, Massachusetts, had built a steam powered crushing mill in the late 1840s. A single stamp unit, the Ball Mill as it was called, was introduced into North Carolina gold mining. By 1857 the mill had evolved into a multiple stamp unit, widely used in the copper region of northern Michigan.[34]

The Pacific steam mill consisted of a battery of two stamps, the downward movement of which was effected by aid of high-pressure steam, while the upward movement was produced by steam at low pressure. The stamp heads were affixed directly to the pistons. Induction and eduction of steam to the cylinder's upper portion, which produced the "strike" of the stamp, was made through a direct communication with the boiler, and could be kept at any desirable pressure. Light weight, only 1,500 pounds without boiler, the machine was intended as a prospecting mill. Including stamps, mortar, and cylinder, it was a mere two feet square by six feet high.[35] A few years later, Pacific redesigned this mill, and marketed it as the

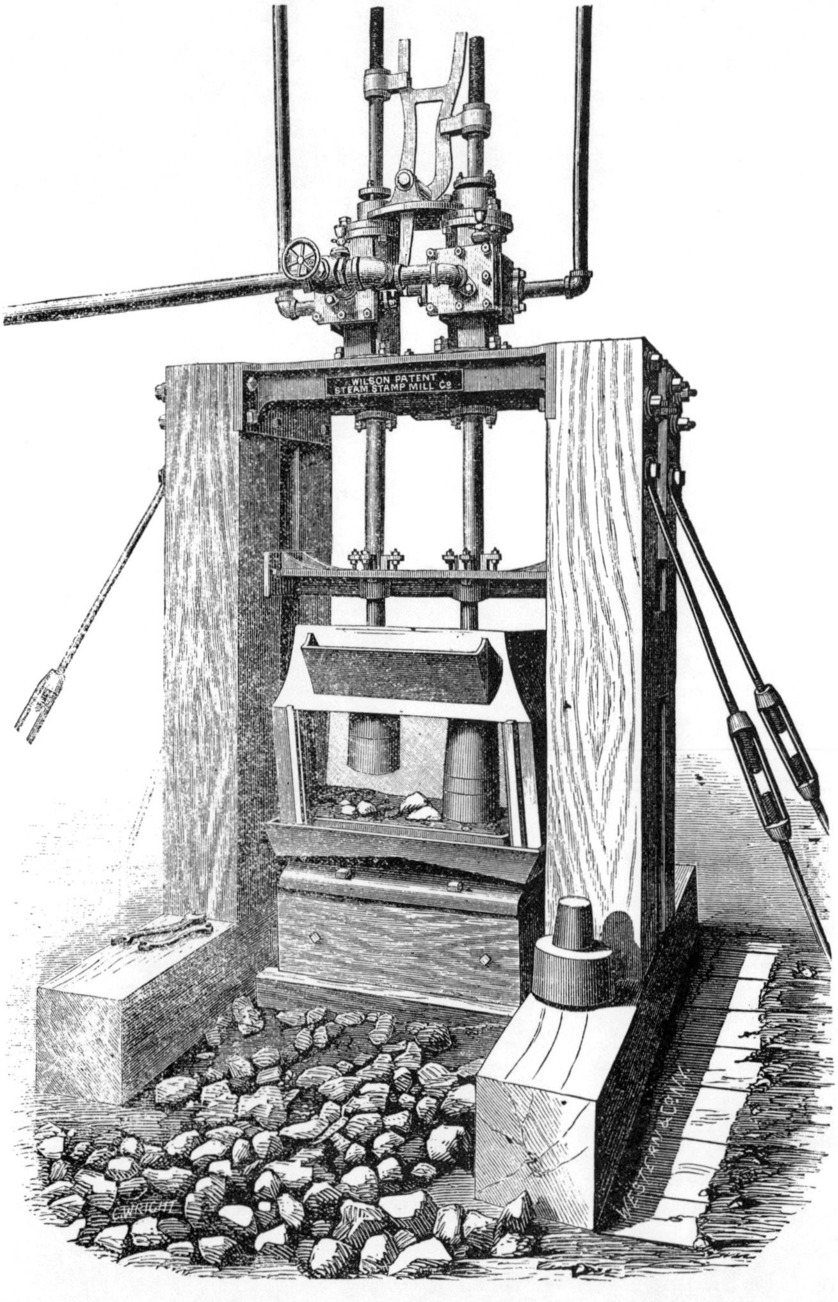

The Wilson Steam Stamp Mill was patented about 1868 by Furman R. Wilson of San Francisco. It ran on the same principle as the Ball Mill, and was touted as the "most durable, efficient and economical stamp mill ever used." It took steam direct, thus eliminating an engine or any intermediate power between it and the boiler. The Wilson Mill could be set up in a couple of days. Manufactured by the Pacific Iron Works, the unit weighed about 6,000 pounds, and was shipped complete with a twenty-horsepower boiler. The mill consumed no more than two and a half cords of wood every twenty-four hours. It was advertised as being able to crush twenty to forty tons per day. (*Mining and Scientific Press*, April 2, 1870)

Wilson Steam Stamp Mill, three of which were constructed in 1870 for the Overman Mine at Virginia City, Nevada.[36]

Construction of mills and Varney pans constituted a large portion of Pacific's mining business in 1867. That spring the company shipped mills to New Mexico, to the Humboldt River, to Idaho, and to Mexico. In 1869 it received an order for mining equipment from Colorado — the first sent to California from that Territory — an indication that West Coast machinery was beginning to be appreciated in the Rocky Mountain region.[37] The company also turned out sugar mills for Tahitian plantations.[38]

Under the superintendency of George W. Fogg, Pacific's foundry was entirely rebuilt in the early 1870s. A three story brick building, "perhaps the best in San Francisco at the time," was erected to house a pattern-making shop, foundry, and machine shop, the latter completely refitted with new tools. The revamping enabled the firm to turn out the heaviest equipment. As one of the largest and best equipped shops on the coast, Pacific enlarged its maritime work, fitting out tags and steamships running in California's bays and rivers. It built one of the first compound engines ever made in San Francisco for the steam tug *Fearless*, running in Coos Bay, Oregon. And about

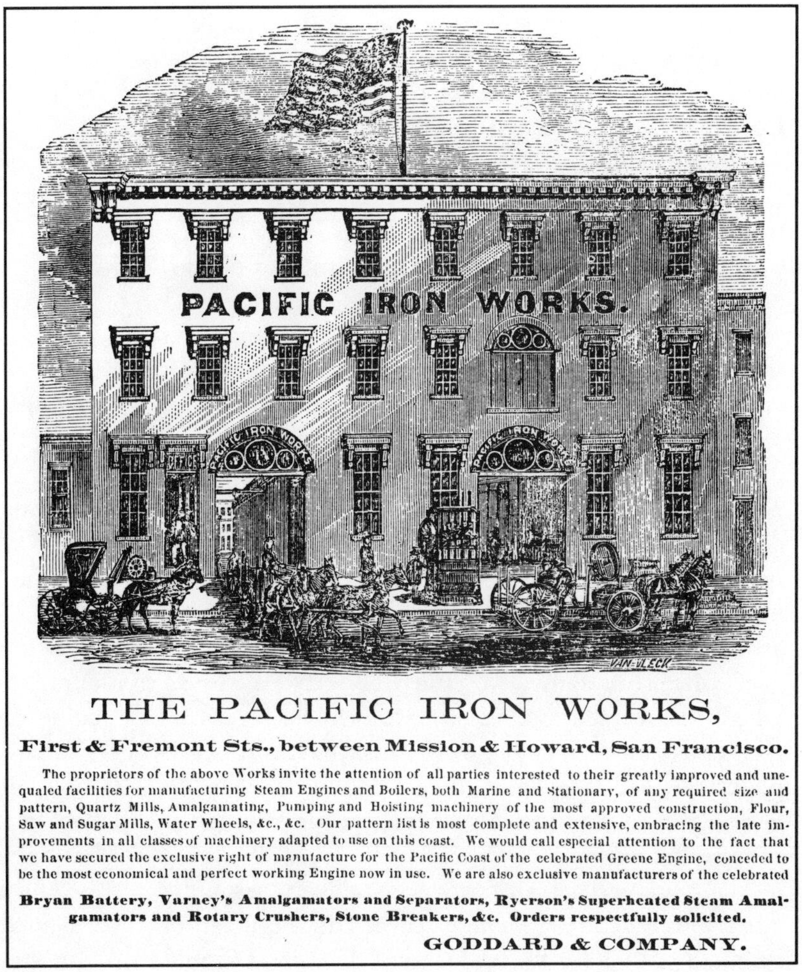

An ad for the Pacific Iron Works that appeared in the *Mining and Scientific Press* on July 1, 1865.

this time Pacific constructed a shallow draught, propeller-driven freight boat to run on California's Owens Lake. Launched in 1872, this vessel carried wood, merchandise, and machinery to the Cerro Gordo Mining District.[39] Manufacture of mining appliances did not lag behind maritime work at Pacific.

In the early 1870s Pacific excelled in the design and building of mills, hoists, and smelters.[40] Much of this equipment was tailored for the environment in which it would be set up. Stamp mills and hoists were designed to be sectional, no piece weighing over 200 pounds, allowing for easy mule transport into rugged terrain. Little wonder, therefore, that mining companies in Mexico and more isolated regions of Western North America sought Pacific's machinery.[41] In 1870 Pacific received an order for machinery from the Lane and Kurtz Cariboo Mining Company in British Columbia: hoisting and sawmill equipment, and two ten-inch Cornish pumps capable of throwing 2,000 gallons per minute.[42]

With each passing year Pacific modified and improved its products; hoisting and pumping engines were increased in size, a sign that mines were going deeper. In March 1873 it built a large pair of hoisting engines for the Black Diamond Coal Mine on Mount Diablo in Contra Costa County, California. These engines had twenty-four inch cylinders and a five-foot stroke. Together, their weight approached ninety tons. The engines were direct-acting, that is without any intermediate gearing, and built on the English plan with the drum placed on the engine shaft. Because both engines worked together, the generally used large gear-wheels were eliminated.[43]

Improved business led to expansion, and by 1874 Pacific employed 225 men turning out mills, hoists, pumps, and engines.[44] Most of its customers were Comstock mines which needed immense hoists and engines capable of working from 2,000 to 4,000 feet.

The first of this equipment, a hoist patterned after works erected for the Mount Diablo coal mines, went to the joint shaft of the Consolidated Virginia and California mines. It was powered by two direct-action Corliss engines placed twenty-four feet apart, each having a twenty-six-inch bore and a six-foot stroke. They were fitted with balance poppet valves and a cross-cutoff. The cylinders were ponderous, weighing

16,000 pounds each, as was the main shaft: some sixteen inches in diameter in the main journals and eighteen inches where the reels were placed. Double reels for flat wire rope were used. The reels commenced at six feet and wound up to fourteen feet in diameter. The pillow blocks weighed 12,000 pounds each, and the bed plates between 16,000 and 17,000 pounds each. The engines were fitted with two steam brakes, one for each reel, and also had independent hand brakes for each reel. The steam brake was so arranged that the brakeman or engineer could put it on at will, and if the cage should rise higher than it ought to, steam was automatically shut off from the main engines, and the steam brake applied itself, barring any possibility of an accident. Globe valves were used to disconnect any engine at any time. The clutches which threw the engines into gear weighed 4,500 pounds each; having wrought iron bands shrunk around them for additional strength. There were no gear wheels, and the two flywheels acted as brake wheels. Six tubular boilers, fifty-four inches in diameter and six feet in length, furnished steam. Each boiler was set independently, the intention being to use only four at a time; the others kept for emergency or for relief purposes.

The two overhead sheeve wheels, eleven feet in diameter, weighed 6,000 pounds each. The specification originally called for twelve feet, but the railroad could not accommodate that size. This hoist, and a duplicate for the Lady Washington Mine, were the largest erected on the Comstock up to that time. Their weight required they be securely anchored on foundations cut into solid granite.[45]

Pacific shipped the C. & C. hoist in late June 1875.[46] A year later, it put together (in ninety days) the pumping machinery for the Savage Mine, also on the Comstock. According to the *Mining and Scientific Press*, this hydraulic equipment "was the most complete and extensive ever made on this coast." It had a direct-acting compound, none-condensing engine, designed by William H. Patton, consulting engineer of the Savage Mine. Forty feet in length, the engine weighed nearly 100 tons. The engine connected directly with the pump bob at the head of the shaft, without the usual intervening crank and gearing. Valves connected with the Davy differential valve gear, could be operated by an independent engine, which could vary the strokes of the pump without any reference to the motion of the main engine.

The main engine consisted of two horizontal cylinders. The first or initial cylinder was twenty-seven inches in diameter and had an eight-foot stroke. The second expansion cylinder was forty-eight inches in diameter and also had an eight-foot stroke. The two cylinders were placed in line and connected by one piston rod, by which the pistons of both cylinders were operated. Attached to the opposite end of the large cylinder was the bed plate with slides and crosshead, to which was attached the connecting rod, which connected directly to the pump bob.

Steam entered the first cylinder at an average pressure of ninety pounds per square inch, and exhausted into the large cylinder, where it expanded down to a pressure of not more than five or six pounds, thus utilizing the whole expansive power of steam (the great secret behind the economy of the compound engine). A condenser was attached to the engine, which allowed for an increase in power should the shaft be sunk to greater depth. This machinery allowed for a lift of 2,500 feet, but was capable of lifting 4,000 feet, with a capacity of 30,000 gallons per hour. Other pumps were stationed every 250 feet in the shaft, each being supplied from a tank filled by the pump below. The sinking pump, which followed the workings of the shaft, had a twenty-foot stroke, and weighed twelve tons. It was strong enough to withstand the heaviest blast.

The pump rod for this ponderous Cornish engine was made of fourteen-inch square timber, braced with iron straps, and extended to the bottom of the shaft. The enormous weight of this rod was counter-balanced with bobs placed every 500 feet along its length. The engine, with its line of pumps, bobs, and miscellaneous equipment, weighed over 500 tons, and cost in excess of $400,000.[47]

To accommodate this type of construction, Pacific Iron Works enlarged its plant and added new equipment. Just in time, for contracts for other types of machinery poured in: from the Hawaiian Islands, for sugar refining equipment; and from around the Pacific Rim, for marine engines. Early in 1878 the company received an order for an 80-stamp mill from Henry Meiggs and his English associates, owners of the famous Cerro de Pasco Mines lying at 14,000 feet altitude in the Peruvian Andes. The mill, together with pans, settlers, and Blake crushers, was water powered by Leffel turbines. Because this machinery, all 530 tons of it, had to be transported over rough terrain, it was designed to be sectional, no component weighing more than 300 pounds.[48]

While Pacific was perfecting hoists and milling equipment, it also developed a line of furnaces. Working with John Howell, at the Comanche and Indian Queen mines at Benton, Mono County, California, the foundry fabricated a rotary furnace for chloridizing silver ores and sulphurizing copper, gold, lead, and zinc ores. The Howell Rotary Furnace, as it was called, was no more than a revolving tube mounted on carrying wheels, with one end opening into an upper furnace chamber, the other into a lower furnace into which ore was discharged from the tube after treatment. The tube was made in sections having flanges, the upper section being smaller than the lower one, and connected by a mid-section of intermediate size.[49]

Pacific also experimented with copper-lined furnaces, cooling them by applying an iron-jacket around the tuyere and forcing air into the intervening space. The company must have had some success with these air-cooled furnaces, for in 1878 it marketed them to the Eureka Consolidated Company at Eureka, Nevada, and to W. H. Raymond of the Lodi District, also in Nevada. Another furnace was installed at the Modoc Mining Company of Inyo, California.[50]

Pacific did not hit upon a commercially successful smelter until its mechanics designed a water-jacket furnace. That story started in 1868 at Cerro Gordo, Inyo County, California, where Egbert Judson, M. W. Belshaw, and A. B. Elder had been having trouble with their lead smelter. Increased production called for larger charges of fuel, something furnace walls could not stand. Ira P. Rankin remedied the situation by putting a steel jacket around the tuyere and pumping water into the interstice. It worked; no melt downs of furnace walls; increased production. While it is claimed this was the first application of a water-jacket to a furnace in North America, it was not a novel idea.[51]

John Williams, a Welsh metallurgist trained at Swansea, introduced the first water-jacket furnace in 1852 at Drontheim, Norway. His idea was not entirely original either. He merely lengthened the sump of the old German "cupola" furnace, altering it into a well detached from the furnace into which fusionable material constantly flowed. Around this he placed an iron envelope. Between that and the furnace wall water circulated. Addition of a water-jacket, together with a cupola with cast-iron boshes, made the furnace virtually indestructible and eliminated the use of firebrick in smelter construction. The latter innovation was especially important to development of refining in the American West, where lack of a transportation network rendered cost of refractory building materials exceedingly high. John Williams further modified his water-jacket furnace.

In 1858 Williams brought his family to the United States to engage in mining and smelting, first in Connecticut, then in the copper district of northern Michigan. Sometime between 1860 and 1867 Williams erected a small reduction works at Lac-La-Belle, Keweenaw County, where he applied his water-jacket furnace. The rush to promote the West's mineral wealth drew John and his sons to Utah in 1869. Four years later Lewis, John's eldest son, applied the water-jacket furnace to argentiferous lead ores from the Williams and Flagstaff mines in Little Cottonwood Canyon, and the Winnamuck Mine in the Bingham District. About the same time, he modified the furnace for copper carbonate from the Copperopolis Mine, also in the Bingham District.[52]

Simplicity of design, long life, and the ease of smelting silicious ores, were attributes Ira P. Rankin wanted to incorporate into his furnaces. And sometime in 1878 or '79 he and John Williams got together and designed an efficient, portable, and cheap water-jacket furnace. Pacific manufactured it in both 30- and 40-ton models.

The Pacific Furnace, sometimes called the Rankin-Brayton Furnace, was an instant success, recognized for its efficiency and ease of transportation and set up. The unit required no brick work and was ready for use upon arrival. Accepted first in Nevada in 1880 to work lead ore, the furnace was tried on copper ore shortly thereafter. By September of that year, Pacific had shipped three furnaces to Arizona: one for the San Xavier Company near Tucson, another for a New York company operating near Gila Bend, and one for the Copper Queen Mine at Bisbee. The latter mine ordered a second furnace in early 1881. The Huachuca Copper Company at Silver Bell also purchased a 30-ton furnace in early 1881.[53]

The water-jacket furnace made the Pacific Iron Works the "headquarters for smelting equipment."[54] By 1882 fifty of these furnaces were working in Western mining camps, and in fall of that year eleven more were shipped. How efficient these furnaces were is attested to in a letter to the manufacturers by W. P. Miller, superintendent of the Castle Dome Smelting Works, dated February 18, 1882:

> . . . we have two of your 40-ton galena smelters in our works that have been running most of the time for the past two years. With ordinary care I have no difficulty in keeping them in condition for constant service. The patent circulating plates, recently introduced into your water jackets, I regard as the most important improvement ever made in water jacket furnaces, and worth more than the entire cost of an ordinary jacket.
>
> There is no question but that a jacket, for either galena or copper ores, with this improvement, will last two or three times as long as without it, besides lessening the liability to injury and the consequent expense and loss

of time for repairs. Since the adoption of this improvement, I have had no trouble whatever with my jackets. The circulation of the water is so rapid that the inner surfaces of the plates are kept scoured and entirely free from any incrustation, and it seems to afford perfect protection from the effects of mineralized water, and from the overheating and burning out so common in all other jackets. Your style of sectional water jackets have many and obvious advantages over all other forms. The distribution of the blast and the angle of the tuyeres are such as are calculated to secure perfect combustion and the greatest economy of fuel.

With a long and varied experience in smelting ores from a wide range of mineral country, I unhesitatingly give your furnaces the preference over any I have ever seen.[55]

To meet the wants of mining properties in early stages of development, Pacific Iron Works built complete, portable hoists like the one above. They were made with a single reel attached to the shaft by a clutch, and fitted with a friction brake so that the reel could be lowered with the engine stopped. The engine was fitted with a link motion, so it could be worked for both hoisting and lowering. The hoist was geared about six to one. Material and workmanship was advertised as first class, all parts subject to heavy strain being of extra heavy iron.

So successful were Pacific's water-jacket furnaces and other equipment that A. P. Brayton went East to establish sales offices. In late 1882 he concluded an "arrangement" with the Excelsior Iron Works, a pioneer Chicago foundry, for construction of mining, milling, and smelting equipment. Now Eastern purchasers could more easily secure improved Pacific Coast machinery and save on freight charges. The Pacific Iron Works Chicago office, in charge of Willis G. Dodd, was located at 182 Dearborn Street. At the same time, the company opened a New York City office at 35 Broadway, under the charge of H. B. Murray.[56]

Due to development of lead-silver lodes and the opening of refractory copper deposits, the Pacific Iron Works did better than most San Francisco foundries during the early 1880s. It equipped some of the West's most famous mines and reduction works. In 1880 it provided a 20-stamp mill for the Boston & Arizona Mining Company at Contention City in the Tomb-

A Pacific Iron Works single drum hoist for round wire rope. On this particular hoist the drum is loose on the shaft, brought into action by a clutch sliding on two steel feathers set in the shaft. The spur gear was keyed fast to the shaft, and a friction hand brake attached to the flywheel on crank shaft. The hoist was made in two sizes: one for working to a depth of 400 feet, the other good to 600 feet. Both were adaptable to vertical or incline shafts. (*Mining and Scientific Press*, February 19, 1881)

A new pattern double drum geared hoist introduced by the Pacific Iron Works in 1882. The engines were designed for hard service, having extra broad wearing surfaces and girder frames. The hoist was provided with post brakes to each drum, friction brakes on engine shaft, the lever for operating which, together with those for operating the links, clutches and throttle valve, being placed so as to be conveniently handled by the engineer. This hoist was capable of raising a working load of one gross ton of ore, in addition to weight of cable, cage and car, from a depth of 1,200 feet at a speed of 600 feet per minute. The cage furnished with this hoist was fabricated entirely of wrought iron and steel, and had safety clutches. (*Mining and Scientific Press*, February 18, 1882)

stone District. This mill was touted as being the most "complete and expensive yet constructed for the Territory."[57] The following year Pacific shipped three hoists to as many mines at Butte, Montana.[58] It also installed a hoist at the Golden Eagle Company at Globe, Arizona; a pan mill and roasting furnace for the Silver King Mine in Pinal County, Arizona; and a 40-ton furnace for the El Dorado Canyon mill in Arizona, and a long list of orders for lesser machinery was filled for mining companies throughout the Pacific Coast.[59]

Deciding to penetrate the small metal producer market, Pacific down-sized its 30-ton copper furnaces in 1883. Ten-ton furnaces were shipped to Battle Mountain, Nevada, and to Chihuahua and Sonora. The 30-ton units were still manufactured, however. In 1884 two went to Australia, and another one was set up at the Copperosity Mines in Arizona.[60] About this time mining companies began demanding larger units, a sign that ore grade and values were changing. A 60-ton plant was constructed for the New Jersey and Sonora Copper Company at Nacozari, Sonora.[61]

Only six water-jacket furnaces were produced in 1885: three for Mexico and three for Australia. With slacking of smelter sales, Pacific focused again on traditional milling equipment, most of which went to Arizona mining companies: a 10-stamp gold mill with pumping machinery, and 9,000 feet of steel pipe, went to the Southern Belle Mining Company; a 5-stamp mill for the J. H. Berger Mine; and pans and settlers for the Silver King. A large number of Duncan concentrators were sent to Mexico, Central America, and to various mines in California, Arizona, and Washington. Mills were apparently reequipping with more up-to-date machinery.[62]

When demonetization brought a decline in silver mining Pacific turned its attention to manufacturing gold mills. In spring 1886 the company took an interest in the Frisbee-Lucop Mill, a device used for some time in the East to pulverize phosphates. This cen-

Above: Pacific Iron Work's Wheelock variable cut-off engine. This particular engine was constructed for the Boston & Arizona Company mill at Contention City, Arizona. The cylinder was twenty inches in diameter, with a stroke of forty-two inches. Of Corliss design, the Wheelock engine was first exhibited at Paris in 1878, and out performed both European and American competitors, carrying off honors for economy, ease of action and simplicity of its valve gear. According to its manufacturers, the "chief peculiarity of this engine, as contrasted with others having oscillating valves, is that they are in effect balanced, being supported on trunnions or spindles, that relieve the valve seats from undue pressure and permit a movement almost devoid of friction." (*Mining and Scientific Press*, September 25, 1880)

trifugal roller mill, which exerted a pressure of 1,800 to 2,000 pounds, was deemed excellent for breaking quartz and put on the market.[63] Traditional gravity stamp were pushed hard in Latin America. Units with Duncan concentrators were sold to the Santa Lucia Company of Honduras, and the Santiago Company of Columbia. Several 10-stamp mills were fabricated for mines on the Mother Lode in Siskiyou and Tuolumne counties.[64]

In 1887 Pacific began experimenting with electrification of mining machinery. On March 26 the *Mining and Scientific Press* reported the company "making some appliances for the new process of Dr. Rae, for utilizing electricity in the prevention of loss of mercury and amalgam in silver mining on the Comstock."[65] Like many things in mining, Jules H. Rae's process was not new. He had been experimenting with electrolysis since the mid-1860s, obtaining patents in 1867, '68, '69 and 1872. Apparently the success he attained by 1887 was the result of a series of failures which yielded nothing but experience.

Trials in 1887, however, satisfied both the inventor and the Pacific Iron Works, and the Rae process was offered to the mining community. In late spring Pacific applied the Rae principal to working of ores and tailings at the Douglas Mill in Dayton, Nevada. A dynamo (motor) was set up to operate seven settlers. Not only did electrical current power the settlers, anodes were set in their bottoms to attract amalgam. It was claimed that electrical current passing between the anodes not only saved bullion but restored floury mercury to a globular form. Working 135 tons of ore per day, the mill claimed a four to ten percent increase in bullion yield, and $5,000 a month in mercury retrival. That summer the process was put to work in the Christy Mill at Silver Reef, Utah.[66]

Although Pacific introduced a cone ore crusher in early 1888, it chose to stay with more traditional types of batteries. That was unfortunate, for this cone crusher, patented by P. W. Gates in 1881, was revolutionary. As Otis E. Young, Jr., points out, this crusher "was continuous and rotary, rather than discontinuous and reciprocating in its action." Within a massive iron receptacle, hung point down, a massive iron cone, powered by an eccentric gear joined to the power train. Ore was pulverized when the cone wobbled slightly off center. Although the Gates ore crusher, in a more sophisticated version, formed the backbone of modern low grade copper production, Pacific continued manufacturing the Reliance iron frame portable battery, a small machine with large crushing capacity, having a discharge on all sides. The Reliance was made in two models: one with 750 pound stamps, and a prospecting battery with 500 pound stamps. Both could be used on wet or dry gold or silver ores, and were made in sections for easy mule transportation. The mill could be set up in several days after foundations had been completed. Pacific sold more than a hundred Reliance mills to Mexico, Central and South America, and other countries where lumber was in short supply, or where timber framed batteries were subject to shrinkage.[67]

Competition from Eastern manufacturers came with the linking of East to West by transcontinental railroads, and the late 1880s and early 1890s were hard years for all San Francisco foundries. No longer were Western mining districts their exclusive turf. Then too, the mining frontier was expanding. Eastern manufacturers could better serve Sonora and Chihuahua, and interior Mexico. San Francisco companies had to contend with high freight charges and stiff tariffs. Rather than continue in what was perceived as a loosing struggle, some foundrymen choose to retire or change direction.

In May 1890, after thirty-five years with the Pacific Iron Works, A. P. Brayton left the firm. He did not retire, however, but associated himself with the Pelton Water Wheel Company.[68] It was a harbinger of things to come. As related earlier, in late January 1892 the Pacific Iron Works was absorbed by the Union Iron Works, which retained the company's old facilities for construction of smelters, mills, and hoists, as well as a center for job and repair work.[69]

Acquisition of the Pacific Iron Works gave Union a valuable line of machinery: the Pacific water-jacket furnace, the White-Howell ore furnace, the Dodd water wheel, and a variety of hoists, engines, and stamp mills. Included in the acquisition were all of Pacific's patents and patterns, and a first-class plant

The Pacific Copper Smelter represented an improved style of water-jacket smelter. Introduced in 1881, it embraced many features not found in other furnaces, the most notable being great economy and capacity for continuous work. With ordinary care, this smelter could run for six to twelve months with little or no interruptions. Repairs were few and far between. This smelter was shipped complete, ready to set up; an obvious advantage over furnaces that required brick or stone work. Another advantage was its distribution of blast, which avoided dead centers, thus preventing the tendency to freeze. Near perfect combustion assured great fuel economy, and produced a clean, high grade bullion, or matte.

Large numbers of these furnaces operated in the West, the most notable being the Copper Queen works at Bisbee, Arizona. With one thirty-ton smelter, that company turned out in six months, 1,200 tons of blister copper, averaging ninety-seven percent fine. Pacific Iron Works claimed that with favorable fuel and water conditions, these furnaces could profitably work ores averaging four to five percent copper. (*Mining and Scientific Press*, June 18, 1881)

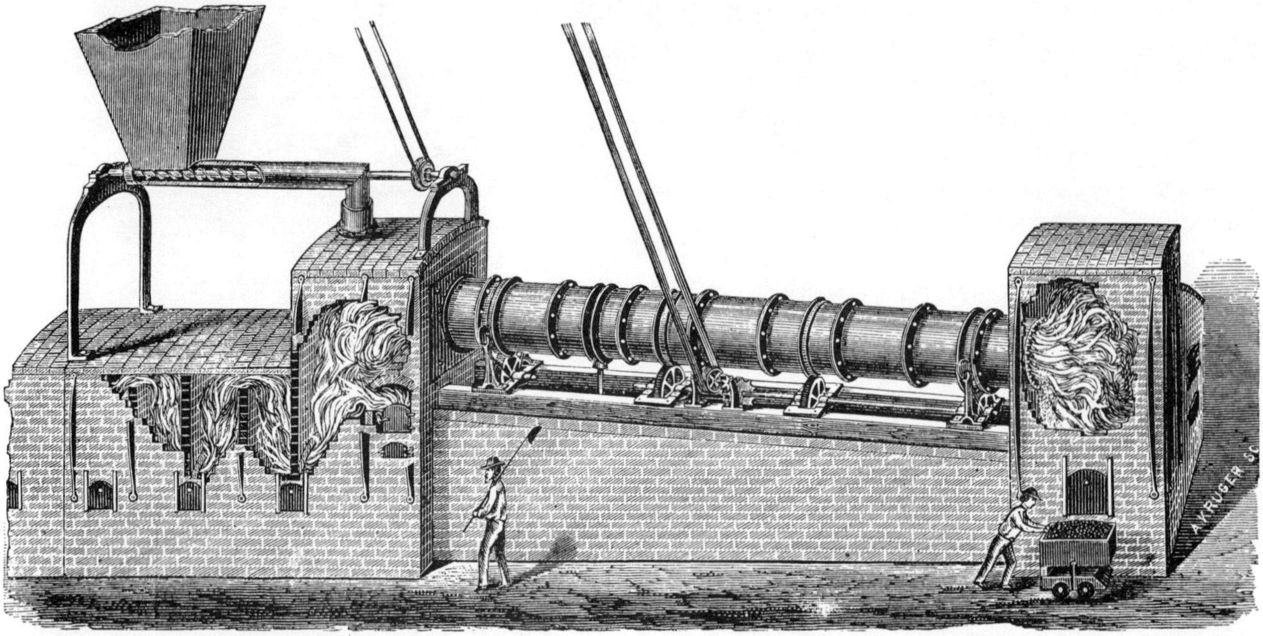

The Howell-White Roasting Furnace, manufactured by the Pacific Iron Works, was conceived to treat "rebellious" ores of gold and silver. Its evolution spanned about eight years. The main segment of the device, the open and inclined revolving brick-lined cylinder, operating with one-fire, was patented in 1864 by G. W. White. The following year, he added an "auxiliary" fire at the lower end of the cylinder, thus correcting a "fatal defect" of exposing ore to too much heat at the beginning instead of the end of the process. This prevented the carrying back of finer ore particles by the draft.

Because heat on the ore was constantly decreased during the process instead of increased, the device was absurd and impractical. After further experimentation, John Howell improved the open inclined cylinder (covered by White's first patent) for roasting and chloridization ores, by merely switching the fires. In 1877 he located the principal fire at the lower end of the tube, instead of the upper end, which increased instead of decreased the temperature. He also located the dust chamber near the upper end, and provided an "auxiliary" fireplace which connected with the flue leading from the upper end of the cylinder to the dust chamber, which prevented the carrying back of finer particles of ore by the draft. This dust, which could amount to twenty percent of the ore processed, was collected and roasted in dust chambers shown in the left of the engraving.

The Pacific Iron Works built Howell-White furnaces in different sizes, having capacity for roasting ore crushed by 5-, 10- or 20-stamp mills. They were easily constructed and required no special skill to operate. They were economical, requiring no more than one and a half horsepower, and from one to two cords of wood per day for the largest unit, which probably accounted for their popularity. By 1878 Howell furnaces were being used to roast and chloridize ore at mines in California, Nevada, and Arizona. (*Mining and Scientific Press*, April 6, 1878)

A Pacific water-jacket smelter for argentiferous lead ore in full running order; the blast furnished by a Baker blower. Powered by an independent small engine, the crusher was of sufficient capacity to crush in a few hours all ore and flux required for the entire day. A horizontal, tabular boiler furnished steam for engines and pumps.

The smelter's water-jacket was made of wrought iron in two or four sections, according to size, so that any of them could be removed without disturbing the rest of the furnace, making all parts accessible for cleaning and repairs. The whole structure was made complete at the shop — mostly of wrought iron, to insure strength and lightness, and in sections, to facilitate transportation and set-up. No brickwork was required except for crucible and two courses on top of jacket, making the cost of erection less than half that of any other style of smelter. A boiler-iron foundation and curb was provided, inside of which the whole structure was built. This prevented escape of bullion in case of leakage from the crucible, which often occurred.

The lead well into which metal ran, was located outside the jacket. It was kept hot by a small fire, the dross and other impurities skimmed off before molten metal was ladled into molds. Made in the jacket, and surrounded by circulating water, the tuyeres never burned out. The jackets were oval in form with round corners, which prevented accumulation of slag, which tended to bridge the ore and prevent it from coming in contact with the blast.

The smelters were made in ten, twenty, thirty, and forty ton capacity models. (*Mining and Scientific Press*, March 14, 1891)

The above illustration depicts a complete Pacific smelting plant of 100-ton daily capacity, designed for large mines or custom works. It consisted of two large water-jacket furnaces with telescope stacks; the dust and fumes passing into a stone or brick condensing flue through connections under the charging floor. Blast for both furnaces was supplied by one large Baker blower, driven from the engine-shaft.

A large Blake crusher, situated above the charging floor, crushed the ore and fluxes, as well as matte and slag, which was often passed back through the furnaces. The slag and matte was raised to the crusher floor by means of a self-dumping skit (traveling on the incline shown in the center of the cut) extending to the top of the crusher floor, operated by a small friction hoist. An elevator (shown to the right of first furnace) provided access to the charging floor for transferring tools and product from one floor to another.

The entire plant was driven by one large automatic cut-off engine, the steam for which was supplied by two boilers. These boilers were set independently, each being of sufficient capacity to run the engine. It was considered best to have one in reserve in case of accidents, such as disconnection of steam pipes, or in the event of cleaning or repair.

A water-supply tank, situated in rear of the engine room, supplied coolant to the jackets. Prior to smelting, rebellious ores containing large quantities of sulphur, zinc, and antimony, were treated in long roasting furnaces housed in a separate building at the side of the works.

The method of arrangement shown here was not arbitrary, as the ground selected for the smelter site often required changes. This plant arrangement was the designed by the Pacific Iron Works, but they often altered their plans to accommodate the requirements of their customers.

At 1891 this smelting plant, the furnace of which is shown at right, was considered one of the most improved, embodying twenty years of refinement. It was a pattern used in large custom smelting works in the United States. The jackets were made of the best fire-box flange iron, in four sections, constructed in such a manner that no rivets were exposed to fire. The tuyeres were made part of the jacket and surrounded by circulating water. Hand-holes with plates and crabs were provided for sediment removal.

A boiler-iron foundation was provided, in which sat the bullion crucible. Such arrangement prevented loss should the crucible crack or leak. The lead well, connected by syphon-tap, was located outside the curb. It too was enclosed. Two slag-spouts were provided for discharge of slag (one at each end of the furnace), which could be easily detached and replaced should burn-out occur.

The superstructure of brickwork was supported by four heavy iron columns, carrying a deck plate made of wrought-iron channel beams, well secured with bolts and wrought-iron plates. The brickwork was securely bound by wrought-iron bolts and heavy cast-iron corner

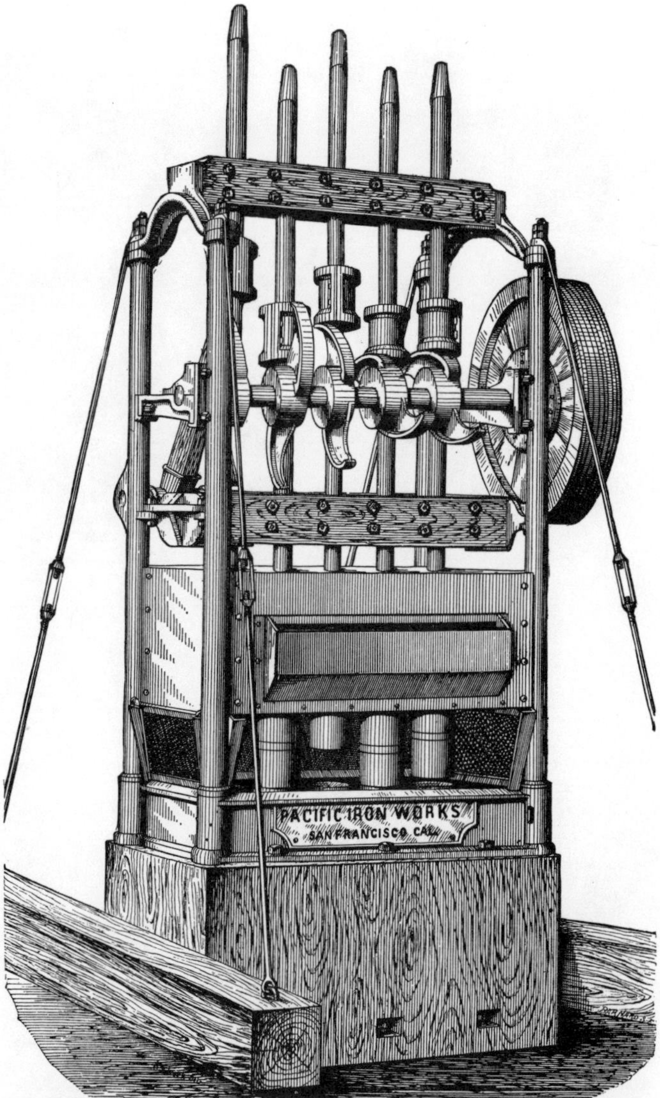

plates, thus preventing any tendency to crack. The brickwork was capped by heavy cast-iron plate, to which were fastened the feed-hopper. Dust and fumes passed into the main condensing chamber through the square pipe at the rear of furnace, shown in the engraving.

A galvanized wind or "bustle" pipe surrounding the furnace, received the blast from the blower, and distributed it through the tuyere pipes. A telescope stack with hood was suspended from proper supports above the furnace. This could be raised or lowered as occasion dictated. The furnace was furnished complete with all pipes and fittings necessary to supply and discharge the water for cooling the jackets. (*Mining and Scientific Press*, February 14, 1891)

While the stamp battery came to be recognized as the standard appliance for crushing quartz ores, it still had detractions. The principal objection lodged against the stamp mill was the expense incurred in erection, on account of millwright work in framing and construction, which could be considerable in localities devoid of suitable timber. The Reliance Iron Frame Battery, made by the Pacific Iron Works, was intended to obviate such objections. With framework made in sections for easy mule transport, the mill could be erected in a couple of days after foundations had been prepared.

The Reliance Battery was extremely popular, especially in Mexico, Central and South America. Another feature of this mill was that it discharged on all sides of the mortar, thus allowing a large crushing capacity. It was equally effective on gold or silver ores with either wet or dry crushing. It was produced with 750- and 500-pound stamps. (*Mining and Scientific Press*, January 29, 1887)

FIG. 1. THE FRISBEE-LUCOP QUARTZ MILL.

The Frisbee-Lucop Quartz Mill was developed by Gideon Frisbee about 1881, and put on the market by Pacific Iron Works in May 1886. It was a centrifugal roller mill, in which two rollers revolved. They were carried around in a vertical plane on the inner periphery of a heavy ring of hard steel, set vertically inside of a casing, against which they exert a centrifugal force of 1,800 to 2,000 pounds each, according to their speed.

FIG. 2. INTERIOR VIEW OF THE FRISBEE-LUCOP MILL.

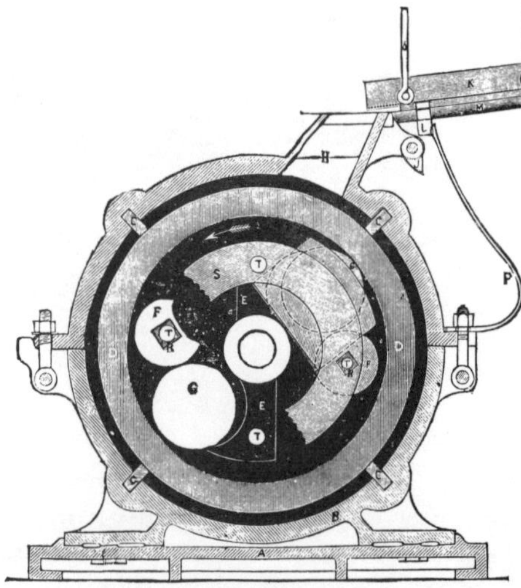

FIG. 3. TRANSVERSE SECTION OF MILL.

The Frisbee-Lucop Quartz Mill was built in three sizes, 24-inch, 20-inch, and a 10-inch, these sizes being the inner diameter of the ring. The 10-inch mill, which had a capacity of 150 to 200 pounds per hour, was intended for assay offices. The 24-inch mill weighed 5,500 pounds, and had a steel main shaft three inches in diameter. A twelve to eighteen horsepower engine drove the mill at 300 revolutions per minute, at which rate the mill would grind a ton of quartz through a 60 mesh screen, and up to two tons of softer material. All ore to be ground was first passed through a crusher to a size of one inch or less, and then fed by a chute directly onto the feed shoe. Cost of the machines were $1,200 for the 24-inch; $800 for the 20-inch; and $400 for the 10-inch. (*Mining and Scientific Press*, May 22, 1886)

specializing in mining equipment. Willis G. Dodd, Pacific's former superintendent, was retained to oversee Union's mining machinery division.[70]

From that time forward a great deal of Union's equipment resembled the old Pacific line. There was no difference in the hoisting engine and a 10-stamp outfit sent off in fall of 1893 to the Harqua Hala Mining Company of Arizona.[71] Although the all-iron 20-stamp mill for the Bardon Mining Company at Guanacavie, Mexico, bore a Union Iron Works plaque, it was identical to those made by Pacific. Union had maintained iron construction in not only its 5-stamp models, but also in 10-, 20-, and 40-stamp units, It even furnished an all-iron 100-stamp mill to the famous Compania Huanchaca de Bolivia in 1894. Design of these mills was so simple that they could be erected with unskilled labor.[72]

The real accomplishment in the mining line came in 1896 when Union constructed two hoists, the largest ever built, for the Anaconda Company of Butte, Montana, designed to sink to a depth of 6,000 feet. The hoists were part of a larger contract which included a refinery with triple expansion Corliss engines rated at 900-horsepower each, three mammoth steam stamps, and several Bruckner furnaces. Thirty-five railroad cars were required to transport this 700 tons of machinery.[73] Identical hoists and engines were also manufactured for the Homestake Mining Company of South Dakota.[74]

With the world crying for copper at 1900, the Union Iron Works employed George Mitchell, a progressive metallurgist who had spent a lifetime studying copper refining, to design and market a large smelter of 300 ton capacity. The main feature of these furnaces were their "adaptability." They were able, in one operation, to reduce sulphide ores to a matte suitable for bessemerizing, without preliminary roasting. These furnaces were immediately accepted by the mining world, with installations at the Copper Queen Consolidated Mining Company, the Greene Consolidated Copper Company, and the Copete Copper Company.[75]

Although shipbuilding may have been Union's mainstay, if not its salvation, it did not turn its back on mining. To house the mining appliance department, the company in 1901 put up a $60,000 three story building on the southeast corner of Second and Folsom streets. The drafting department, pattern shop, administrative offices and show rooms were located there. In the rear of the building, on Folsom Street, was the machine shop, a 100 x 150 foot steel and corrugated iron building surrounded by a gallery twenty-five feet in width.[76]

Mining equipment established the Union Iron Works and made it possible for the company to make the transition to shipbuilding. Shipbuilding, however, ultimately caused the demise of its mining appliance line. Just as they had done in the mining equipment field, Eastern capitalists eyed West Coast shipbuilding, and in 1905 Bethlehem Steel Company purchased the Union Iron Works. The company name died hard, however. For decades the old works were known as the Union yards, and it produced naval vessels during both World Wars.[77]

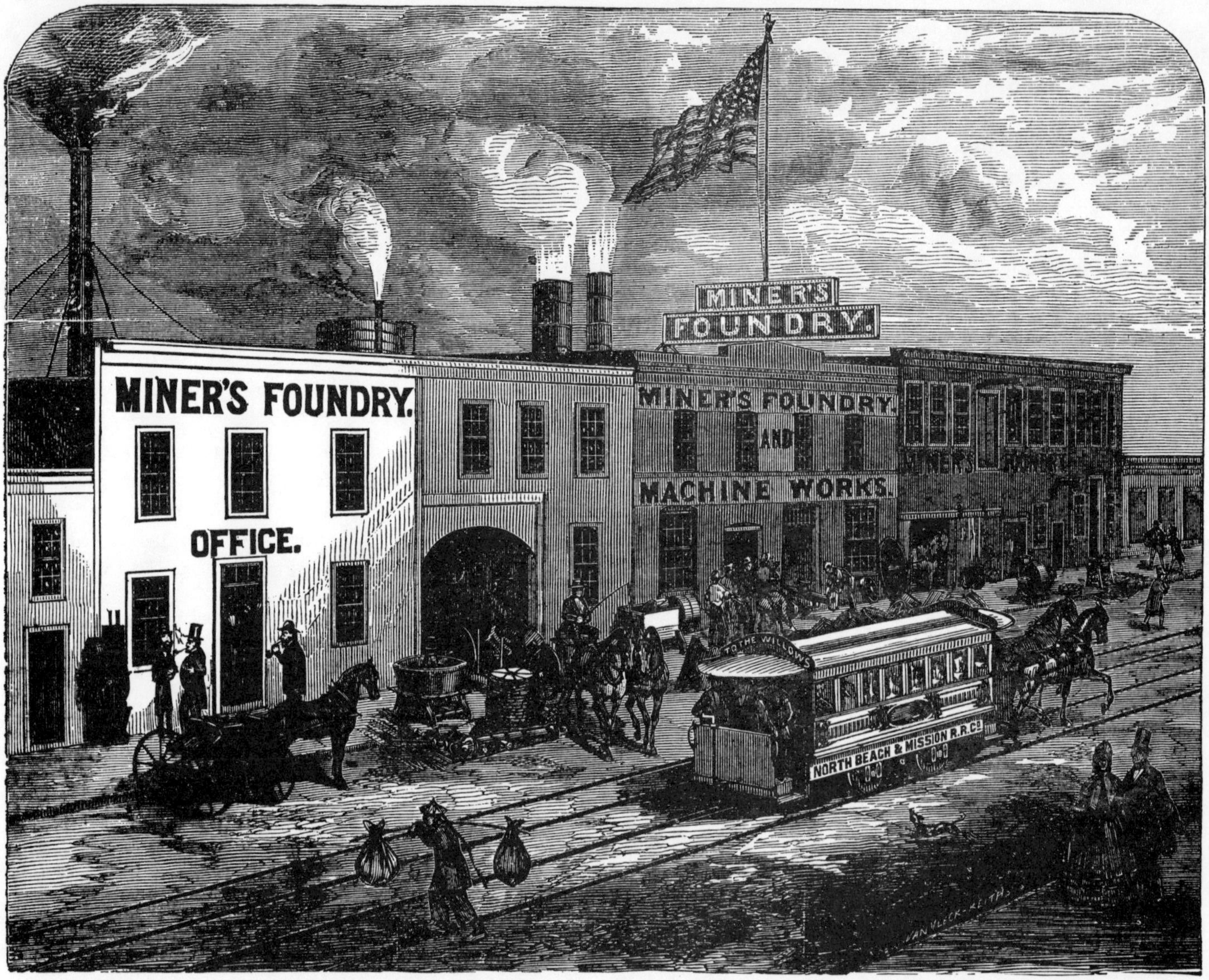

The Miner's Foundry and Machine Works in March of 1865. The firm commenced operations in December 1859, in the center building shown in the above engraving. Their tools consisted of three second-hand lathes and one drilling machine, all driven by a steam engine, the cylinder of which was only four inches in diameter. The company grew rapidly with development of the Comstock. By 1865 the foundry was issuing a sixty-four page *Annual Circular*, which not only advertised its products, but contained valuable tables of distances, and rules for estimating power of steam engines and water wheels. The brochure also included articles appealing to mining men, such as "The Mining Districts on the Pacific Coast," and "Causes of Failure in Mining Operations and their Remedy." (*Mining and Scientific Press*, March 18, 1865)

CHAPTER THREE

The Miner's and Golden State Foundry, Fulton Engineering and Ship-Building Works, and the Risdon Iron Works

THE Miner's Foundry, which built much of the equipment for early Washoe mills, was located on First Street near Folsom. The creation of three mechanics — William H. Howland, H. B. Angell, and James King — the business commenced operation in December 1856. Starting out modestly with three second-hand lathes and one drilling machine, all driven by a small steam engine with a four-inch cylinder, the company grew rapidly, expanding in spring of 1862 and again in 1863. By 1864 the company had more than 150 employees and its capital stock ranged between $150,000 to $200,000. In spring of 1865 the foundry was running twenty-three lathes and eleven drill presses, slotting and planing machines. The foundry was equipped with twelve forges and two cupolas capable of melting twenty tons of iron per day.[1] All of which gave the Miner's Foundry the ability to manufacture and erect some of the finest and heaviest machinery on the West Coast.

Unlike some of its competitors, this foundry did not start out repairing maritime components. Sugar mills was its specialty, and its market was far off in the Pacific. Prior to 1856 planters of the Sandwich Islands (the Hawaiian Islands of today) had to procure heavy equipment either in Scotland or from manufacturers in the Eastern United States. A much closer source came into being with opening of the Miner's Foundry. Mills could be obtained in San Francisco cheaper than in Europe or on the East Coast.[2]

It was not all sugar mills, however. There are design similarities in all types of mills, whether for flour, sugar, or rock, and it was not long before the foundry was manufacturing mills to process gold and silver ores. In 1858 it produced the Howland Rotary Quartz Mill, which consisted of a battery of eight stamps placed in a circular fashion. Weighing 6,000 pounds, and utilizing 500 pound stamps, the Howland Mill was first put to work on Comstock ore by Almarin B. Paul. The appliance was an immediate success and changed forever the methods of reducing silver ores.[3]

The Miner's Foundry worked closely with Almarin Paul and others to alter the Freiberg barrel for Yankee needs. The Howland and Wheeler amalgamators, introduced to miners at Austin, Nevada, was the result.[4] The company also constructed the Hopkins' "Novelty" Rotary Cylinder Grinder and Amalgamator, and the Hepburn and Peterson Amalgamators.[5] In a few years the Miner's Foundry built a solid reputation for turning out large quartz mills, those with engines over 120 horsepower, capable of driving 40-stamps. By May 1865 it was not unusual for the company to have nearly a dozen such mills under construction at any one time.[6]

The Miner's Foundry serviced well the Western mining market. Throughout the early 1870s its stamp mills could be found throughout California's Mother Lode, in Washoe, Nevada, and in Utah. It introduced to the Pike's Peak region Howell's Rotary Crusher, a

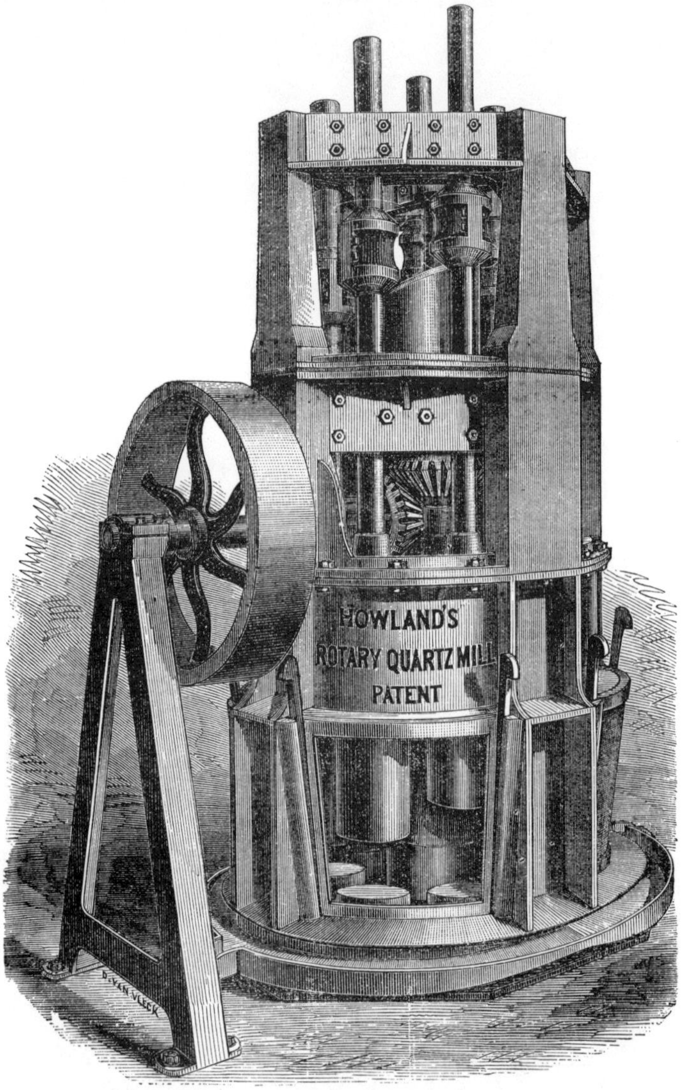

The circular design of the Howland Rotary Quartz Mill (at left) was heralded as a progressive step in the construction of the stamp mill, in that it eliminated the ricketiness inherent in such mills. Design was simple, consisting of a central revolving shaft, which drove a bevel gear seen in the middle, having two triangular or wedge-shaped arm attachments, which elevated and dropped the stamps in turn, two (opposite ones) at a time. The mill was fed from the center, the rock shoveled in, or brought on automatically. Put to work crushing Comstock ore by Almarin Paul, the Howland Battery stood the test of time, working for seven years at his Gold Hill Mill. The Howland was also introduced in the Grant District, east of White Pine, about 1869. The mill came in two models: and 8-stamp unit, weighing 12,000 pounds, costing $1,800; and a 12-stamp unit, weighing 18,000 pounds, costing $2,800. (*Mining and Scientific Press*, March 6, 1869)

device whose rotary beaters "driven at immense velocity," reduced quartz to an "impalpable powder." Then, as now, advertising men were skilled in concocting sales pitches.[7]

In late February 1867, while at the heights of its production, the Miner's Foundry experienced a fire, probably incendiary in origin, which destroyed most of the firm's pattern shop. Although damage was estimated at only $18,000, much of which was insured, the loss was nevertheless devastating. Patterns constitute an important part of a foundry's inventory, replaced only at great expense and expenditure of time.[8] The foundry recovered, however, and devoted the remainder of that year building steam engines, boilers, and pump components for other foundries. Of the latter, was an order placed by the Union Iron Works, for a large number of sections for Cornish pumps destined for the combined shaft of the Imperial and Empire mines on the Comstock.[9] At the same time, the Miner's Foundry undertook production of the Tangential Turbine Water Wheel which found wide application on the Mother Lode where there were fast moving streams.[10]

In late 1868 the foundry experienced labor problems. Rather than face a strike, the company closed. Farsightedly, the company found that the "only true solution to the antagonistic questions between labor and capital" was under a "cooperative plan," and re-

At left: The Tangential Turbine Water Wheel, manufacured by the Miner's Foundry, provided power for many California quartz mills. The wheel was eight feet in diameter and cost $500. With sufficient water velocity (a head of ninety-four inches, generated by an eighty foot fall), this turbine wheel could drive a 10-stamp mill. (*Mining and Scientific Press*, April 14, 1866)

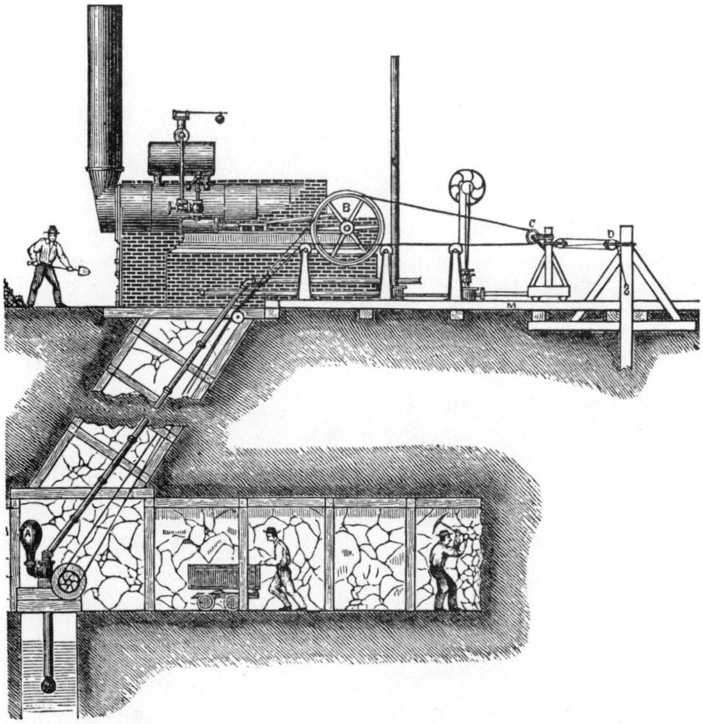

The Goodwin Pump, manufactured by the Miner's Foundry, was designed for both hydraulic mining and fire engine application. It was made in sizes to throw from five to 600 miners' inches of water per minute.

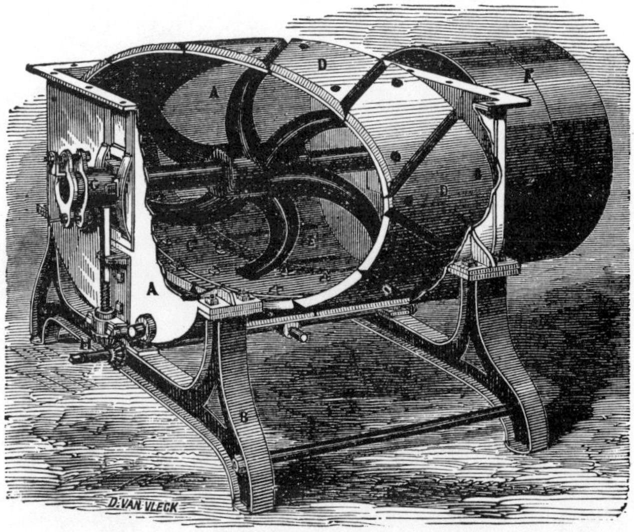

The Novelty Amalgamator, another Miner's Foundry product introduced about 1865, was nothing more than an horizontal cylinder. It came in various sizes, ranging from twenty to 1,200 pound capacity. A model was kept constantly running at the foundry so that prospective buyers could evaluate its worth. (*Mining and Scientific Press*, January 21, 1865)

The illustration at upper left shows the Goodwin Pump as applied to deep mining. Power generated by a steam engine was transmitted to the pump via wire rope and adjustable grip pulleys. *A* represents the pump in an incline (though it was equally well adapted to vertical shafts) connected with the driving pulley, *B*, of the steam engine, by a wire rope which passed over the extending pulley, *C*. Because the pulley was supported by a frame, resting on track, *M*, it could be easily moved. As the pump was lowered into the mine, the pulley slid forward, eliminating the necessity of lengthening the rope each time the pump was lowered. The pulley, *D*, tightened the rope whenever it became slack, or after the pump was set.

opened six months later. This cooperative plan allowed laborers to share in the profits "over and above a fair interest on the necessary investment."[11]

Despite a sluggish economy in 1869-70, the Miner's Foundry reported "getting a fair amount of what was being done." This included construction of four Severance and Holt's Diamond Drills, with fifteen horsepower portable engines, destined for the Harpending Mines of New Mexico, and the Black Diamond Coal Mine at Mount Diablo.[12]

With advent of deep mining in the early 1870s, the Miners' Foundry, like all the others, added pumps to its line of equipment, principally the Goodwin and West Pump, a reciprocating cylindrical pump capable of throwing 1,200 miners' inches of water per minute. This pump was used for both hydraulic mining and water control in drifts and deep shafts.[13]

Despite a "fair amount" of equipment business during the mid-1870s, which included filling orders for mines in Mexico, Peru, and Chile, the Miner's Foundry could not compete with larger San Francisco companies. To remedy that situation, the firm merged with the Golden State Foundry in summer of 1877; the consolidation resulting in the formation of a new company called the Golden State and Miners' Iron Works. Before looking at this entity let us backtrack and say a little about the beginnings of the Golden State Foundry.[14]

The origin of the Golden State Foundry lay with George K. Gluyas, who was previously connected with the Aspinwall Steam Transportation Company which spanned the Isthmus of Panama with a railroad. Gluyas established himself in San Francisco in 1849, and a year later opened the Sutter Iron Works on the corner of Beale and Folsom streets. For nine years he manufactured a variety of items: sawmills, grist mills, and machinery for steamboats such as boilers and engines. Although he turned out some quartz mills, maritime work was Gluyas's specialty, and the Sutter Iron Works produced the machinery for the *Captain Sutter*, the first sternwheel steamboat built on the West Coast.

In 1859 Gluyas moved his foundry to a new location at 19 and 21 First Street, near Market.[15] Two years later, he was bought out by Palmer, Hanscom & Company, owners of the Eagle Foundry on the corner of Battery and Bush streets. Two brothers, I. W. and R. F. Knox, machinists with a knack for engineering amalgamation equipment, entered the picture about this time and bought out Hanscom. The new firm of Palmer, Knox & Company set up the Golden State Foundry and moved its entire plant, stock, and tools to the building formerly occupied by Gluyas on First Street.[16]

In 1858 the Knox brothers invented and patented the Knox Amalgamation Pan, one of the earliest improvements introduced into the business of quartz mining, and the first of a class of machinery which revolutionized gold and silver mining. In varied forms, these pans found wide acceptance in California as well as in the amalgamating rooms of quartz mills worldwide. The popularity of the Knox Pan propelled the Golden State Foundry to the forefront of quartz machinery manufacture by 1867.

The firm broadened its equipment line, turning out stamp mills, hoisting engines, Wheeler and Randall pans, and Knox and Osborne quicksilver furnaces, the latter widely used in California's mercury production. Mining appliances were not the only concern of the Golden State Foundry, however. In 1865 it employed forty men and consumed four tons of pig iron daily preparing tresses for bridges spanning that portion of the Western Pacific Railroad between San Jose and Stockton.[17] At the same time, it turned out castings for mills at Virginia City, as well as pumping works and hoisting machinery for Pine Grove.[18]

A fire in 1870 momentarily halted Golden State Foundry work. It apparently recovered rapidly, for the *Mining and Scientific Press* commented that the firm "seem to have been very fortunate in securing work since their late fire." The following year a large quantity of Stevenson pans and settlers were turned out, and their Knox pans went to developing gold mines in Plumas and San Diego counties.[19] By spring of 1877 Golden State was manufacturing Frue concentrators to work gold-bearing beach sands in Oregon and Washington.[20]

In summer of 1877 the land upon which the Golden State Foundry had been situated for fifteen years was purchased by Huntington, Hopkins & Company, another mining equipment manufacturer. Rather than close, Palmer, Knox & Company combined with the Miner's Foundry, the new company operating under the name Golden State and Miners' Iron Works. Directors of the new corporation were Cyrus and William Palmer, H. B. Angell, I. W. and R. F. Knox.

The old Miner's Foundry buildings were enlarged and all Golden State machinery moved to that location.[21] The combined company focused on building Frue concentrators, mostly for Arizona. The fact that the Silver King Mine was using four of them, turning out 2,500 pounds of concentrates daily, worth $2,000

The Severance and Holt Diamond Drill was conceived in 1860 by Professor R. Leschot, a French engineer, who found that by arranging diamonds in a ring-shaped bit or boring head (Fig. 1) a core could be cut out of rock rapidly and economically. Messrs. Severance and Holt, practical San Francisco mechanics, obtained Leschot's patent rights and set about improving the drilling machine. With help of the Miner's Foundry, they rendered the diamonds settings more secure, increased the machine's speed, and regulated the feed of the drilling bit to compensate for varying hardness of rock. A swivel head was added which allowed the working of the drill at any angle. In 1870 the machine was introduced to the United States in several styles, the most common being the "No 1 Prospecting Drill"(Fig. 2) which consisted of an upright boiler firmly bolted to a cast iron frame supporting a five to seven horsepower engine. The "No 2 Prospecting Drill" (Fig. 3) was similar to the first, only larger and more powerful. It had a tubular boiler, $3\frac{1}{2}$ by 7 feet, and its double engines produced 12 horsepower. It was geared to twice the speed of the No. 1, and mounted on large wheels with broad felloes for easy transportation in rough mining districts. Advertised as being especially adapted to well-boring, draining and prospecting, the machine produced bores from two to five inches in diameter, as desired, and to any required depth. Total weight was about 3,800 pounds. Both the No. 1 and No. 2 were self-contained units. (*Mining and Scientific Press*, May 7, 1870 and March 11, 1871)

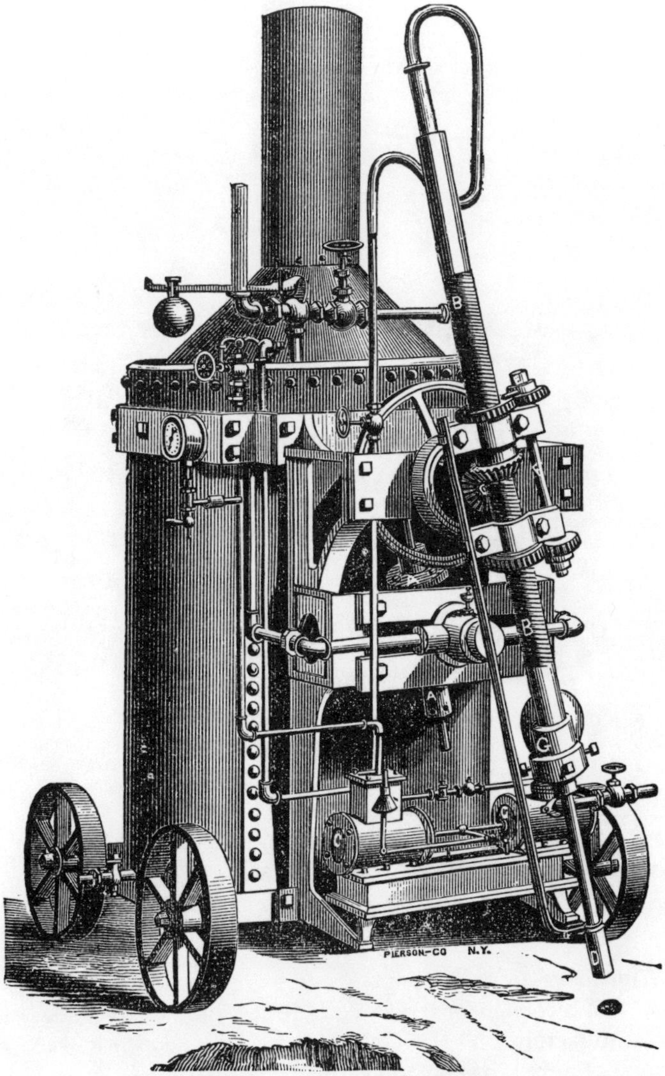

FIG. 2—THE No. 1 PROSPECTING DRILL.

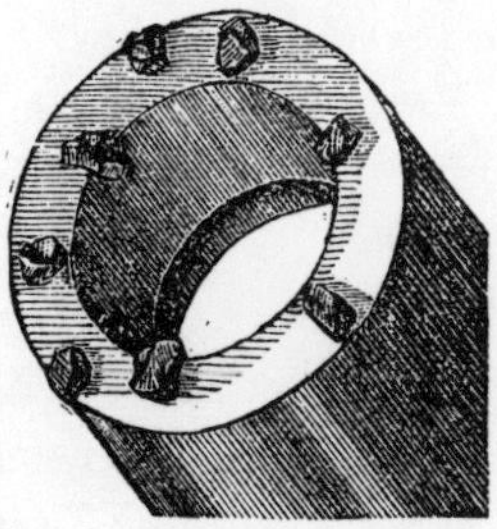

FIG. 1.

FIG. 3—No. 2 PROSPECTING DRILL.

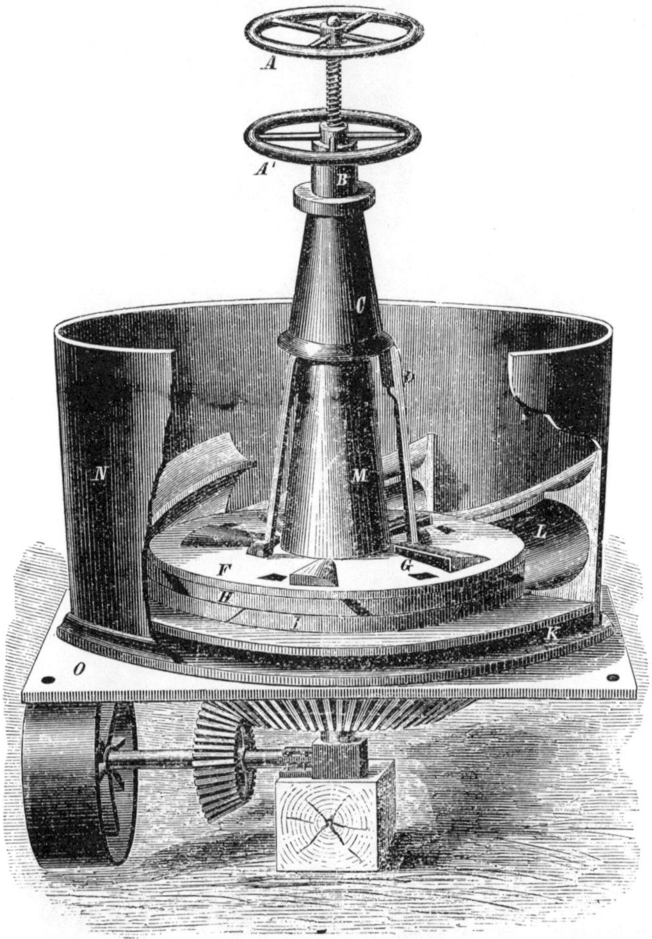

The above pan was designed and patented about 1869 by C. C. Stevenson, of the Douglas Mill, Gold Hill, Nevada. Manufactured by the Golden State Iron Works, it was one of a multitude of devices collectively called Washoe Pans. The difference between it and other pans, was a series of curved grooved guides, L, placed around the side of the pan, which were supposed to keep the pulp level during rotation of the muller, which allowed a more thorough grinding of the mass. This construction also prevented mercury from settling to the bottom of the pan. (*Mining and Scientific Press*, April 2, 1870)

per ton, was the best advertising the foundry could have.[22] As a result, it soon had considerable supply work from the territory, mainly shoes and dies, and other castings for stamp mills. The foundry also performed most of the work for California quicksilver mines, and in fall of 1881 it completed a 20-stamp mill for an Idaho mine and a large double hoist for a mine in Utah. Several single hoists were also fabricated. Added to this, the company turned out equipment for San Francisco's emerging cable car roads.[23]

Although the foundry built numerous stamp mills, and even produced the J. B. Osborne steam wagon, a vehicle used on the Mojave Desert to haul ore from mine to mill, a major turning point occurred in company history in 1882. That year it undertook construction of several large dredges for the Panama Canal.[25]

Weighing more than 300 tons each, these dredges were shipped to Philadelphia where they were installed on big scows, and towed to Central America.[26] Put to work on the canal, the dredges far exceeded expectations, each moving 10,000 cubic yards of debris in a ten hour day. From then on, the Golden State and Miners' Iron Works specialized in the construction of giant suction devices. For nearly two decades their dredges worked clearing and deepening California's harbors and coastal rivers. By 1900 the foundry had abandoned the manufacture of mining equipment altogether. A similar parallel can be found in the history of the Fulton Iron Works, which grew to be the third largest mining equipment manufacturer in San Francisco.

New Yorker David B. Hinckley, a boiler maker by training, was a founder of the Pacific Foundry and Machine Shop in 1850. Five years later he dropped out of that business, for what reason no one knows. Hinckley pooled his assets with those of two other machinists, William E. Worth and William A. Field, and together they purchased a cupola and opened the Fulton Foundry and Iron Works at 45, 47, and 49 First

The Frue Ore Concentrator, manufactured by the Miners' and Golden State Iron Works, won First Premium and a Silver Medal at the San Francisco Industrial Fair of 1880. It out performed other concentrators on the market, saving from fifty to 100 percent more concentrates than vanner machines. Within a year there were 250 Frue machines operating in the West.

Street, between Market and Mission streets. General foundry and architectural work seems to have been their main concern at first, for much of the ornamental castings which adorned early San Francisco buildings came from the company. Like other foundries, Hinckley & Company, as it was known, broadened its operations by manufacturing steam engines and a variety of milling equipment.

By 1864 production of mining appliances formed a large percentage of company work. Besides fabricating custom stamp mills, most of which were sold to California quicksilver mines, Fulton produced Prater Concentrators, and the well-known Brodie Crusher, an invention of James Brodie, foreman of the foundry. These devices effectively treated sulphuret ores and were used by major mining companies of Nevada's Eureka District, as well as in Mexico.[27]

Fulton engines powered most California sawmills. And like so many other foundries, the company engaged in repair and building of marine equipment. In 1864 the company built a poppet valve steamboat engine of the type used on the Mississippi River — the first in California. That engine was quickly accepted by the state's river trade.[28] Two years later, Fulton turned out machinery for a vessel belonging to the Russian Telegraph Expedition and equipped a steamboat for Nelson, Anderson and Runyon Company, then being built at the John North shipyard.

Although Fulton's place on First Street was spacious in comparison to other foundries, heavy production schedules forced a move to larger quarters. In summer of 1867 Fulton acquired property on the northeast corner of Fremont and Tehama streets and erected a new plant. Running through the block from Fremont to Beale Street, the building had a frontage of 111 feet. The main structure was three stories high. Its machine shop was eighty by fifty feet; the blacksmith shop forty feet square, and the molding shop eighty by fifty feet. Fulton added a second, larger cupola with a melting capacity of fifteen tons.[29]

Fulton expanded its business in the early 1870s by introducing Vance's Little Giant Quartz Mill and

N. P. Boss designed special grinding pans for the continuous process. These pans ground pulp produced by the battery, thus relieving the amalgamating pans of that duty. They were made four feet in diameter, and had solid ring shoes and dies slotted for a short distance from their inner edges. The pulp was introduced at the center of the pan, and a rubber gasket between shoe ring and muller-plate, forced the mass between the shoe and die before being discharged. The engraving shows pans receiving the pulp of a 10-stamp mill. One pan placed below the other, received and reground the pulp as it came from the first. These grinding pans were made by the Fulton Iron Works.

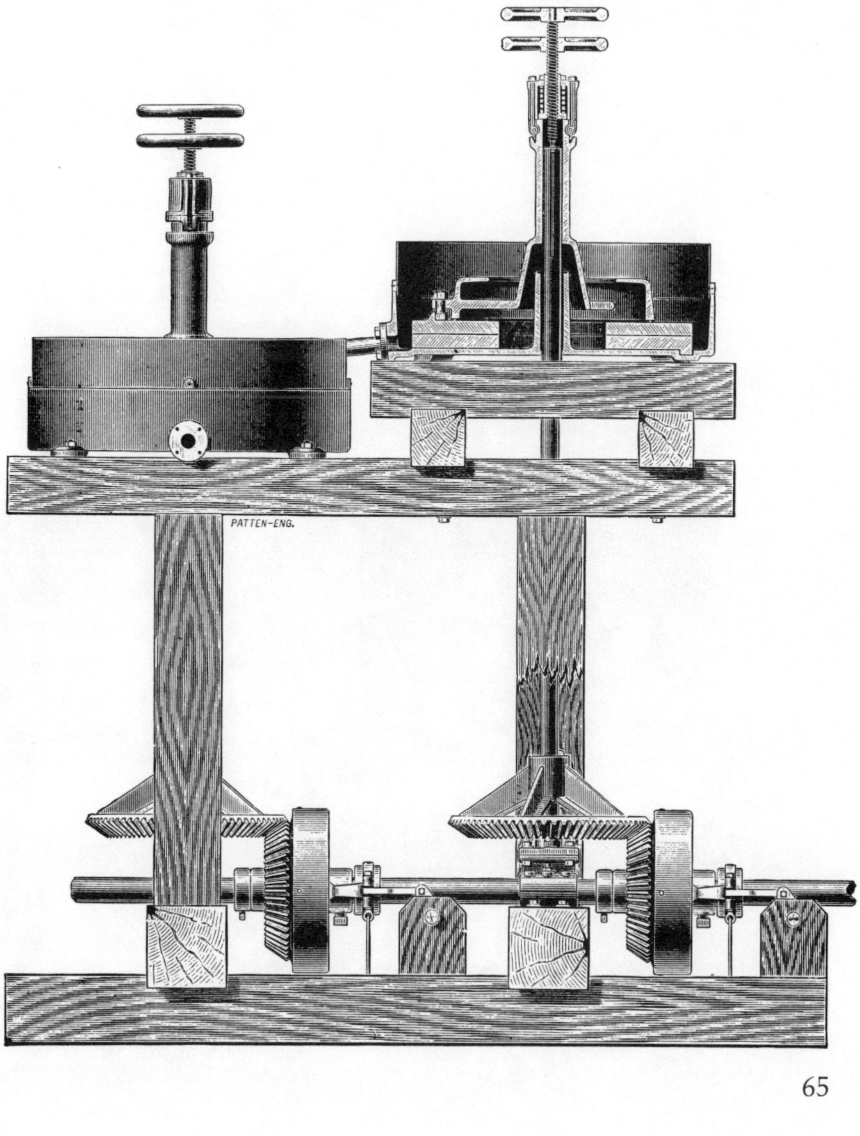

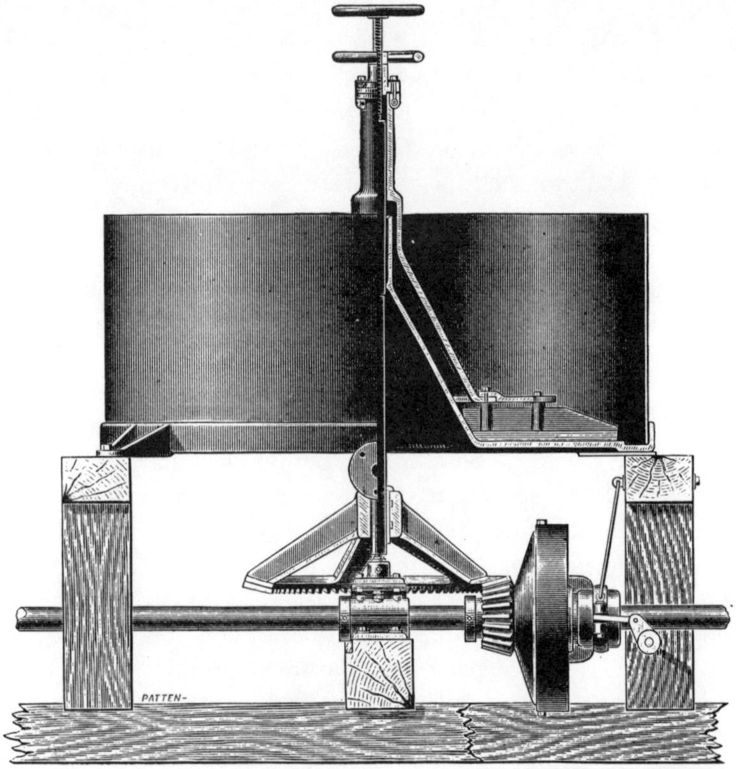

The N. P. Boss Standard Settler, at left, was designed for use in continuous mills. The muller arms and driver were cast in one piece, to which the muller-plate was bolted. The friction gear and spindle step were similar to those on pans. Shoes working close to the bottom, formed a strong undercurrent which swept the bottom, and saved wear and tear. Quicksilver bowls were provided the same as in pans. The settlers were fitted with heavy wrought iron sides and were connected together near their tops, similar to pans. The cone, cast with the settler bottom, was wide at its base, which allowed the shoes to work close to it, thus preventing accumulation of pulp on the bottom which could cause clogging and breakage.

Below: The Boss Standard Amalgamating Pans were made heavier than other pans, and had a steam bottom extending into and around the cone. The cut at bottom left shows the pan in section, with quicksilver bowl and pan siphon in position for draining the pan of mercury and amalgam. These pans had an inside diameter of four feet two inches and weighed 8,500 pounds.

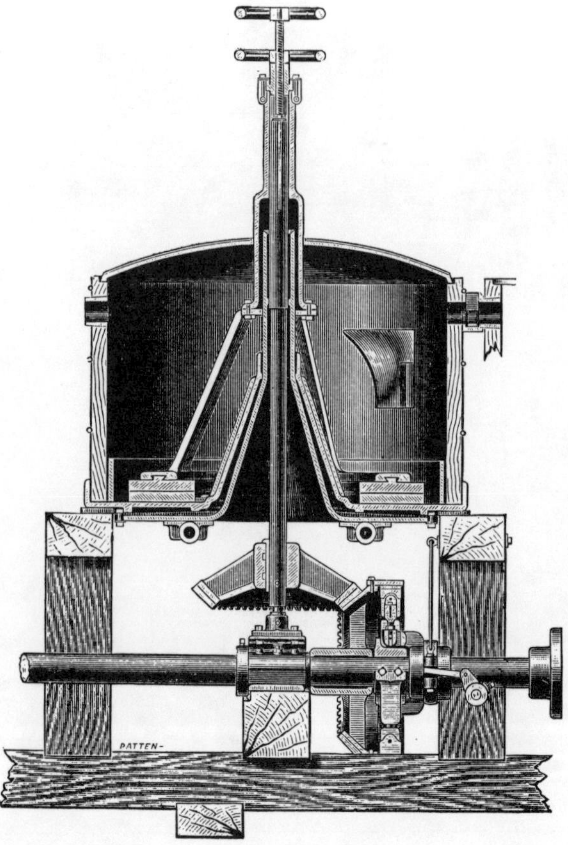

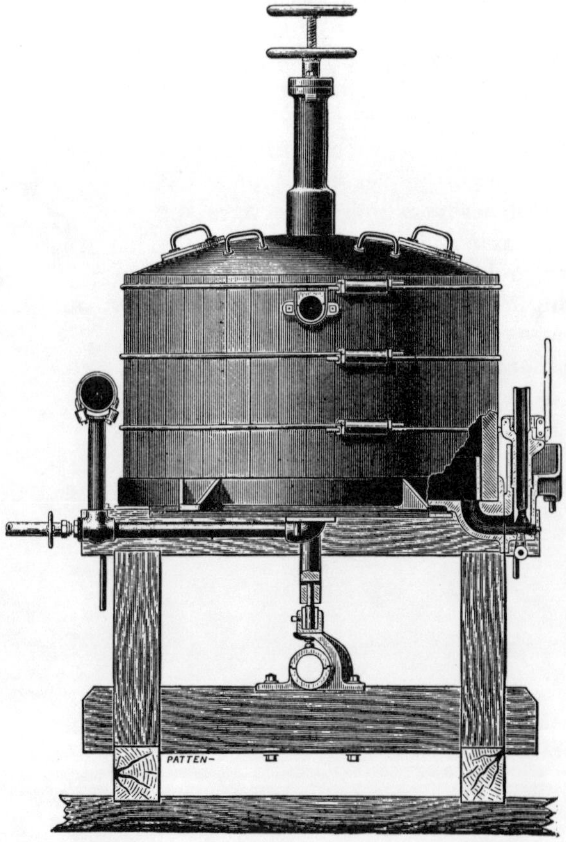

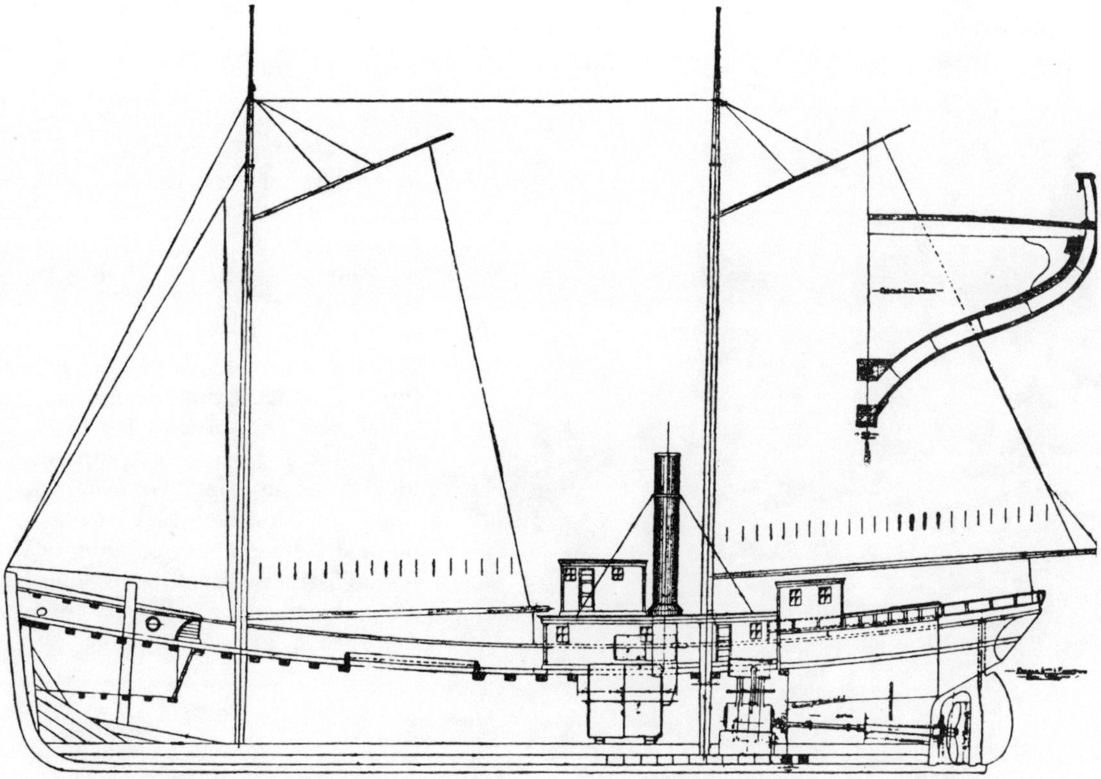

Ever since lumber mills were first built in California, Oregon, and Washington, trade between the lumber ports and California's coastal cities was carried on by a large fleet of sailing schooners. These schooners were built at San Francisco and were designed to withstand rough weather and carry heavy cargoes, most of which was placed on deck. In the 1880s Fulton Iron Works began building these schooners, equipping them with compound engines, and large capacity boilers. (*Mining and Scientific Press*, June 16, 1888)

turning out castings for smelters being erected in Nevada.[30] California quicksilver mines also took many of its hoists and pumps. At the same time, Fulton increased its output of architectural castings. In 1873 it turned out 275 tons of castings for the London and San Francisco Bank, one of the finest buildings erected up to that time in San Francisco.[31]

The company also took a hand in opening the Colorado River to navigation, a far reaching event in the history of Arizona mining. In spring of 1872 Fulton built two engines of sixteen-inch bore and five-foot stroke for the Colorado Steam Navigation Company, the latter organized in 1869.[32]

Although it built stationary and marine engines, from six to 175 horsepower, Fulton was severely handicapped in this field by not being able to manufacture boilers. That situation was remedied in late 1877 when it absorbed the McAffee & Spiers Boiler Works, one of the earliest boiler works on the coast.

Since the late 1860s McAffee & Spiers had constructed and repaired boilers, and they prided themselves in refitting marine engines. In 1872 the firm installed new boilers in the steamer *Cricket*, and built a fire box boiler for a steam dredger at Portland, Oregon. They also furnished tanks for the California Sugar Refinery, as well as flue boilers for mining companies.[33] Merger of Fulton with McAffee & Spiers was a shrewd business decision that created opportunities, for the company turned out a profussion of boilers and machinery for dredges and vessels for the next six years. In 1878 it fitted new boilers to the old steamer *Wilson G. Hunt* and the new steamer *Princess*.

Hinckley, Spiers & Hayes Company, as Fulton was now called, still concentrated on mining equipment. In fall of 1877 it designed a heavy hoist for the Tybo Consolidated Mine in Nevada,[34] and a year later erected a 10-stamp mill with pans, settlers, retorts, and clean-up pans for the Tombstone Mining Company of Arizona. This machinery was powered by the San Pedro River, through a twenty-five inch American turbine wheel. A similar mill, powered by an engine and boilers, was erected at the McMillan Mine, also in Arizona.[35]

Production of stamp mills kept Fulton going through the next year. A 30-stamp mill was turned out for the Noonday Mine at Bodie, which included

The Tustin Ore Mill, made by the Fulton Iron Works, was perfected at the Oro Plata Mine, Murphy's, Calaveras County, California. Its principal parts were two rollers, slotted grate dies, screens, and a yoke bar. The barrel of the machine revolved on two hollow trunnions. Ore delivered to the barrel via hopper and feed tube, fell on and under the rollers, and was crushed between and under them. The pulp dropped through slotted grate dies onto revolving screens which sized the particles. The Tustin Mill was advertised as producing a pulp free of slime, ideal for amalgamation or concentration. It was claimed the mill could save twenty to forty percent more gold and sulphurets. The largest model Tustin weighed 13,000 pounds, and was adaptable to either wet or dry crushing. (*Mining and Scientific Press*, July 24, 1886)

twenty-four 5-foot pans, twelve 8-foot settlers, a 23 x 48-inch engine, three steel boilers; the whole outfit. A similar mill was erected for the Western Mining Company at Tombstone: 20-stamp mill with fourteen 5-foot pans, seven 8-foot settlers, an 18 x 14 engine, two 54 x 16 boilers, and miscellaneous equipment. That was just the beginning of Fulton's Tombstone business.

Besides furnishing Tombstone five mills (the Tombstone, Corbin, Contention, Sunset, and Grand Central), Fulton supplied hoisting engines to most of the district's mines. To supervise erection of this equipment, the company sent in A. J. McCone, vice president of the Fulton Foundry at Virginia City, also owned by the San Francisco company. McCone was assisted by J. P. McAllister, a mechanic who had cut his eye-teeth on the Comstock machinery trade. The latter Scotsman established a foundry in Tombstone and made Cochise County his home for more than a decade.[36]

By the early 1880s Fulton's business had changed considerably. Marine equipment, principally boilers and engines for steam ships engaged in the Hawaiian trade and the Coos Bay lumber traffic, was the company's chief concern.[37] Fulton also constructed dredging equipment for the U.S. government, used on West Coast bays and rivers.[38] The California coasting trade greatly benefited the Fulton Iron Works. In spring 1887 it built ten marine engines for steam schooners plying the lumber trade. As these schooners steamed up the coast, made their landings, and then sailed back with the strong and favorable winds of summer, their engines were used exclusively for auxiliary power. Only in case of head wind, or in making their landings, was steampower used.[39] By the mid-1880s Fulton had definitely relegated its mining equipment business to the miscellaneous work category. Nevertheless, that type of work was considerable.

In spring of 1886 it turned out four large Huntington mills. What really kept its mining equipment department going, was the Tustin Ore Pulverizer, a mill tested at the Oro Plata Mine at Murphy's in Calaveras County, California. That mill apparently competed favorably with gravity stamp mills, at least it was so advertised. In spring of 1886 Fulton shipped many Tustin crushers, included one to a gold mine at St. Petersburg, Russia. At the same time, it fabricated a 20-stamp mill for the Argo Mine in New Mexico, and built an ice plant for Merry, Faull & Company, a San Francisco slaughterhouse.[40] Fulton was indeed versatile.

Despite a lucrative marine trade, Fulton still looked after its mining customers. In summer 1887 a double reel hoist, extremely simple in construction yet very strong, and a 10-stamp battery with all attachments was shipped to the Golden Gate Mill at Lost Basin,

Arizona.[41] Although Fulton designed and erected custom mills anywhere in the world, it still manufactured a standard variety of mills for both gold and silver ore. Updated from year to year, these mills always embodied the most modern equipment.[42]

In mid-October 1887 a disastrous fire swept Fremont Street, destroying much of the Fulton Works, as well as the Eagle Brass Works next door, and the railroad car shops of John Hammond. Fulton lost its foundry, machine shops, and drafting room. Only the office opposite the main works was untouched by the fire. The firm began rebuilding at once, and by December 1887 new quarters on the old site were well underway, and machine tools had been procured from Parke & Lacy, Tatum & Bowen, and H. P. Gregory.[43]

In early June 1892 a second fire swept the block bounded by Howard, Folsom, Fremont and Beale streets, destroying Fulton's entire plant, valued at $300,000. Insurance did not cover the loss and it looked as if the third largest foundry in San Francisco would cease to exist. It did not vanish, for men of means came to its rescue.[44]

That summer Fulton was reorganized and incorporated with $1,000,000 capitalization. The three original stockholders: David B. Hinckley, James Spiers, and Daniel E. Hayes, remained in the business. Four new men entered the scene. The first three, F. A. Huntington, Henry P. Christie, and Richard D. Chandler,

The Fulton Rock Breaker was patterned on the same principle as the Blake, of a central oscillating pitman, transmitting its motion to the moving jaw through toggle-jointed plates. Fulton, however, endeavored to design their machine so that its wearing parts were more accessible for repair and replacement. The general form and design of this rock breaker was such as to insure the greatest possible strength. The above illustration shows the rock breaker with its stationary jaw closed, ready for work. (*Mining and Scientific Press*, February 15, 1890)

The Fulton Iron Works produced high class rolls for ore crushing. Crushing capacity of such rolls depended on their diameter, speed, the space between crushing surfaces, size and hardness of the material fed to them, as well as the regularity of feed. The patent rolls shown in elevation and plan in the engravings on this and the next page were designed to withstand the severest strains imposed in dry crushing of ore. (*Mining and Scientific Press*, March 7, 1891)

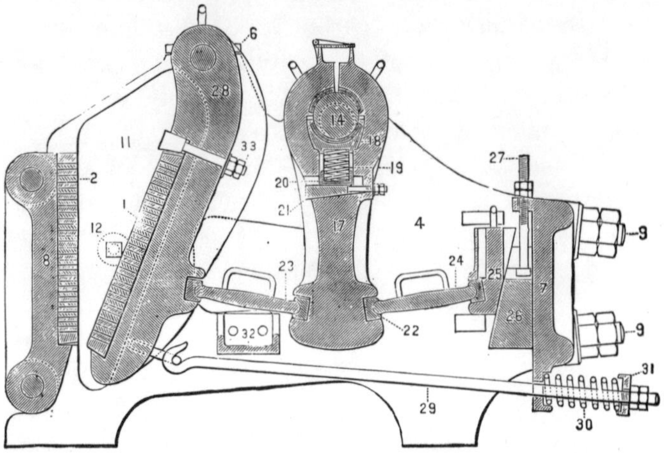

The above illustration shows the Fulton Rock Breaker with jaw open to allow easy removal of worn shoe, die and cheek-plates. The stationary jaw was rigidly held in place by means of flat iron bars having eyes forged on their ends, which slipped over shafts at top and bottom of the jaw. By taking out the pins in the end of the upper shaft, and loosening nuts holding them at back of rock-breaker, the upper bars could be slipped off, allowing the jaw to be pivoted on the lower shaft, as shown.

A fair idea of the construction of the shoe and die may be obtained from this sectional view of the Fulton Rock Breaker. They were composed of alternate layers of wrought iron and hardened machine-steel bars placed on edge and held together by a heavy wrought iron band shrunk around them. Being softer than steel, the iron wore away rapidly, creating a corrugated surface on the shoe and die, effecting better crushing. The surfaces of the iron bars wore a short distance below those of the steel, allowing the hardened steel to do most of the work. There is no danger of the bars becoming loose and falling out, it was claimed, as the crushing of rock forced the wrought-iron bars more firmly within the band.

Dry fine crushing rolls required the highest class of workmanship and material. The roll-shells were fabricated from "perfectly homogeneous" mild steel. For the heaviest work, the roll shafts were made of steel and the rolls driven direct by pulleys without use of gearing. The frame had planed lugs for attaching a slide-rest to true up worn roll-shells.

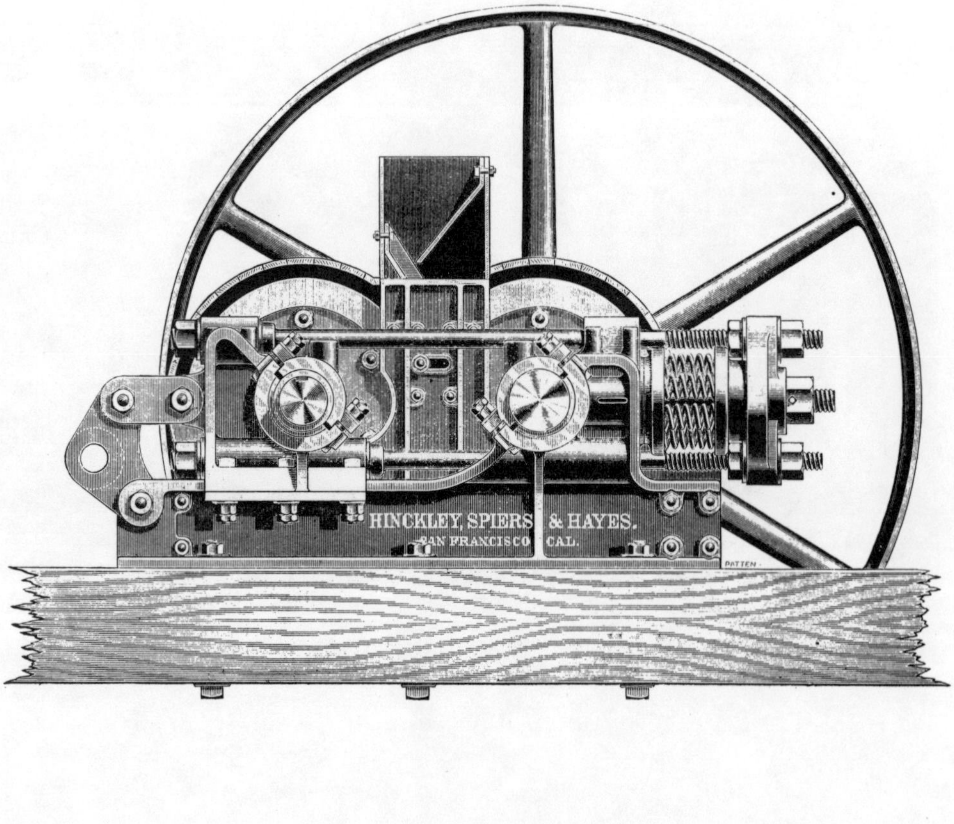

The Fulton Iron Works produced more than engines, hoists, and rock breakers. They fabricated the N. P. Boss Improved Retorts and Bullion Melting Furnaces, which no precious metal mill could be without. Amalgam was placed on trays in the body of the retort, and as quicksilver was vaporized, it passed through the nozzle into the condenser, from which it flowed into a receiving tank. When all the quicksilver was driven off by the heat, the gold and silver was left as a spongy mass ready for melting. That was done in the melting furnace, the gold and silver from the retort being placed in a crucible, melted, and poured into a bullion mold, from which it was taken, cleaned, and made ready for shipment.

Often two or more retorts were placed side-by-side, having flues running to one stack. The bullion melting furnace was usually built into the retort setting both for convenience and to save expense. (*Mining and Scientific Press*, June 14, 1890)

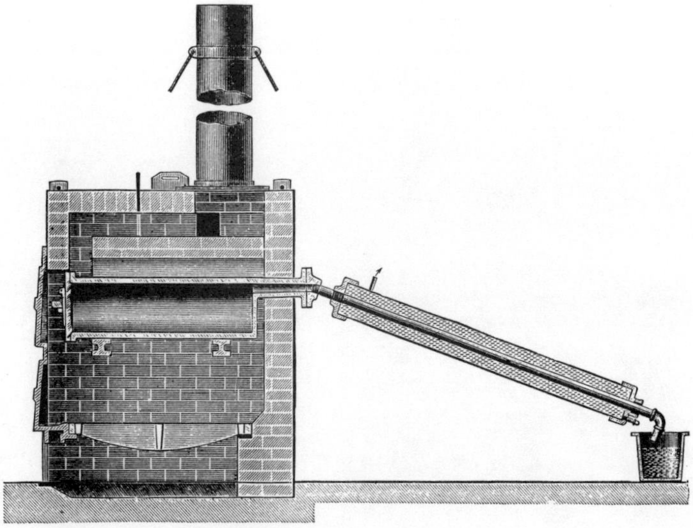

SECTION OF RETORT.

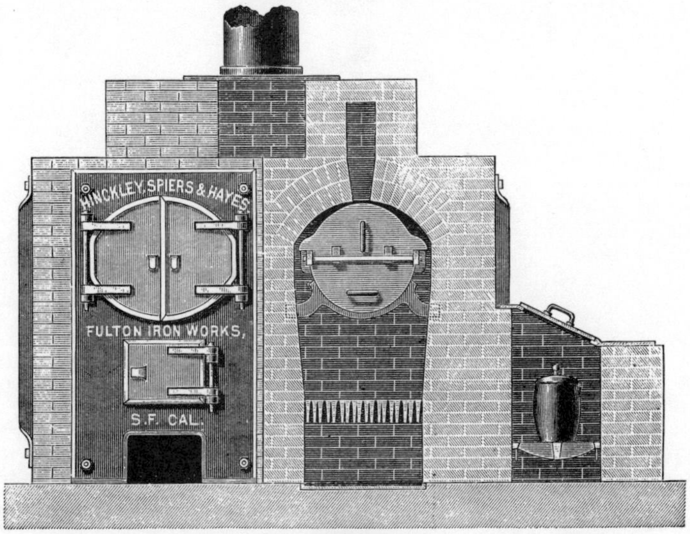

DOUBLE SILVER RETORT AND MELTING FURNACE.

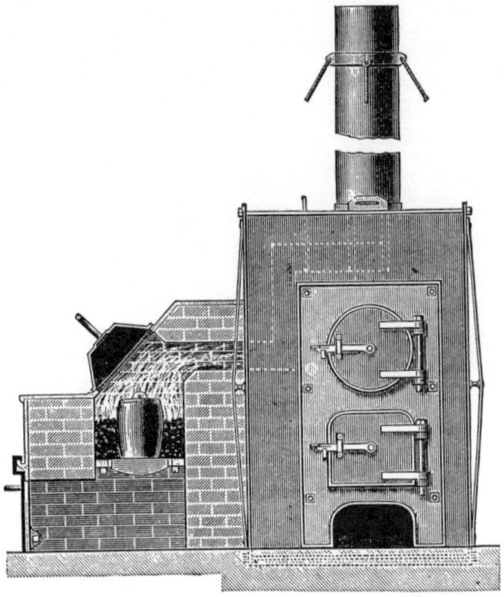

SINGLE RETORT AND MELTING FURNACE.

The N. P. Boss furnace operated on the principle of an ordinary forge. The pan constituting the bottom of the furnace was filled with a mixture of bone-ash and fire-clay, thoroughly tamped down and then scooped out, leaving a lining about two inches thick of the mixture over the entire inner surface of the pan. On top of that, and confined by a wrought-iron basket or grate, charcoal and bullion were placed. As the bullion melted, it percolated through the charcoal to the bottom of the pan, and as it accumulated there it was to a certain extent refined by absorption of the base by the bone-ash lining.

Molten bullion was drawn off into molds by tapping the discharge spout at the bottom of the pan. As soon as a bar was poured, the discharge spout was stopped with a plug of bone-ash until sufficient bullion had accumulated for another bar. After pouring the first bar, succeeding bars were cast at intervals of about fifteen minutes each, bullion and charcoal being piled on top as fast as necessary. One man could easily pour six bars, of about 4,500 troy ounces each, in two hours.

This style of melting had advantages over crucible reduction, in that it was done by a reducing flame instead of an oxidizing one, thus avoiding loss of silver by oxidization.

Two blast tuyeres passed through the water-back of the furnace, and the blast was supplied by a Root blower. A blower of suitable capacity and an extra pan-bottom were supplied with each furnace.

These rolls were particularly adapted to the second crushing of galena and all ores where a gradual reduction in size was advisable.

On a previous page we showed you Fulton's high-class dry-crushing. Above is the company's cheaper crushing rolls. These were chiefly adapted to comparatively coarse work, but could also be used for tolerably fine crushing when the rolls were in good condition and the shells new and of good material. No gearing was used on these rolls, they being driven instead by pulleys on each roll-shaft.

These corrugated rolls gave excellent results, when large quantities of material from a rock-breaker were required to be crushed to a size of about three-quarters of an inch. They also crushed material to the same degree of uniformity as the smooth rolls, with little sliming. The roll-shells were made of mild cast steel for long life. (*Mining and Scientific Press*, April 11, 1891)

were machinery dealers and manufacturers. The fourth, James G. Fair, was the millionaire of Comstock fame. Fair brought to the company capital and land on which to expand.

Fulton's old location was confining, hardly large enough to carry on a marine trade. If it was to survive in that line of enterprise, Fulton needed land with plenty of waterfront. Property near Oakland was sought; the city offering $10,000 to help the company move. But Oakland was unable to provide good title to harbor privilege. James G. Fair had land at Harbor View, on the bay shore of the San Francisco peninsula, and he could provide title to harbor access for five lots of solid ground fronting the bay, having 980 feet of water frontage. Most importantly, water was deep there; no wide mud flats as in the southern part of the peninsula.

In fall 1892 Fulton relocated to Fair's land, and by December the company was hard at work preparing the new site. A first class modern, fireproof plant was built, five times larger than the previous one.[45] The foundry building was completed in February 1893, and core ovens and brass furnaces installed. The machine shop was equipped with machinery of the latest design. In mid-March sixty men went to work; within a few weeks the original work force of 150 employees was back to work.[46]

The first shipbuilding venture of the Fulton Engineering and Ship-Building Works (a new name) was a 255-foot ferry boat for the North Pacific Coast Railway Company for use on San Francisco Bay.[47] Four more contracts for steam vessels and numerous repair orders for Government vessels swamped Fulton the next year. Revival of gold mining following the collapse of silver increased the company's orders for mining equipment. Concentrating plants and small

Another variety of Cornish rollers produced by the Fulton Works. (*Mining and Scientific Press*, April 11, 1891)

The three-stamp mill, fabricated by Fulton Engineering and Shipbuilding Works, was handy for prospecting or when a larger and more pretentious plant could not be secured. The unit was designed to be operated by either horse, steam, or water power. The frame could be taken apart and put together again with no cutting or refitting. The stamps were ninety-drop, and weighed 250 pounds each. The entire outfit as shown, including frame, weighed 4,500 pounds. (*Mining and Scientific Press*, May 25, 1895)

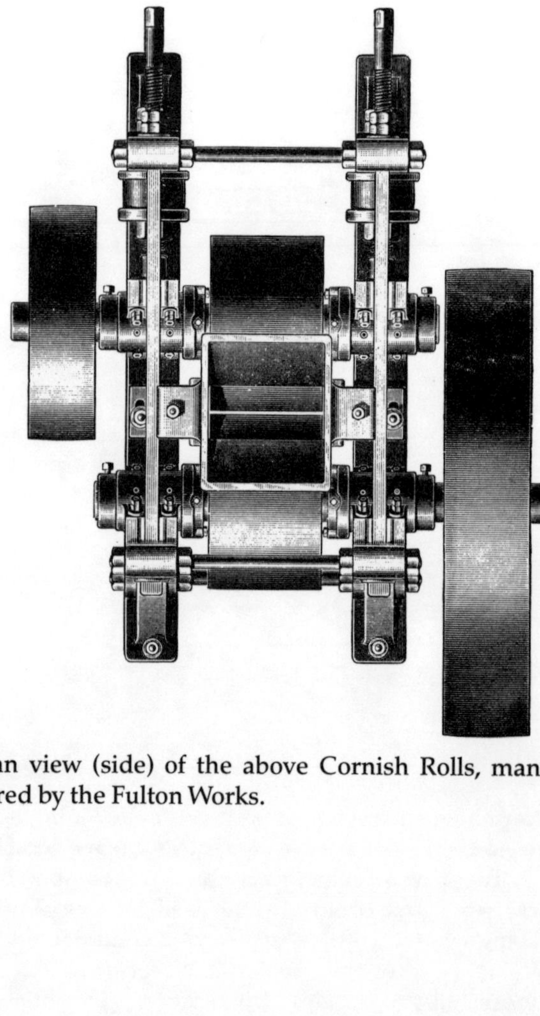

A plan view (side) of the above Cornish Rolls, manufactured by the Fulton Works.

stamp mills were shipped to mines in northern California and Arizona.[48]

Fulton no sooner combated one crisis, than it faced another. In the 1890s electrical power was coming onto the mining scene, rendering steam plants obsolete. Fulton met that threat by shifting to an entirely new field: air transmission. In 1895 the company began turning out compressed air plants, both for military and mining purposes. Its largest plant, a 700 horsepower unit, was designed to operate the pneumatic guns at the San Francisco Presidio. Fulton unveiled an even larger plant, powered by an 800 horsepower duplex Rix tandem compound engine, for the North Star Mining Company of Grass Valley.[49]

Although Fulton Engineering and Ship-Building still advertised itself in 1897 as a supplier of "all sorts of mining and milling equipment," it concentrated on building ocean-going vessels. That year it turned out four schooners, three of them for the lumber trade and one for fishing. It also built a steamer for the California Navigation Company, which plied between San Francisco and Stockton. This vessel had a speed of twenty knots.[50] When company advertisements disappeared from the *Mining and Scientific Press* in 1898 it was obvious that Fulton had abandoned altogether its mining equipment adjunct. Once the third largest producer of mining appliances on the West Coast, Fulton had made the transition entirely to shipbuilding and refitting. Within a few years it vanished from that scene as well, swallowed by Bethlehem Steel.

LEWIS COFFEY & RISDON'S

STEAM BOILER AND SHEET IRON WORKS.

The only exclusively Boiler Making Establishment on the Pacific Coast. Owned and conducted by Practical Boiler Makers. All orders for New Work or the repairing of Old Work, executed as ordered, and warranted as to quality.

Old Stand, corner of Bush and Market Streets.
Opposite Oriental Hotel, San Francisco, Cal.

LEWIS COFFEY, J. N. RISDON

Risdon Iron Works' first ad appeared in the *Mining and Scientific Press* on June 29, 1861.

The Risdon Iron Works was San Francisco's fourth largest manufacturer of mining equipment. Founded about 1852 by Lewis Coffey and J. N. Risdon, the Coffey and Risdon Boiler Works as it was advertised, started on the corner of Bush and Market streets opposite the Oriental Hotel. By 1861 it employed forty to fifty craftsmen.[51] Over the next ten years the company experienced phenomenal growth, branching from boiler making into fabrication of refrigeration and gas-making equipment. In 1872 it built for S. L. Mastick's ice plant the first surface condenser ever made in San Francisco.[52]

The key to Risdon's success, however, was cast and sheet iron pipe for mines and municipal water works. In 1870 the company fabricated twelve miles of thirty-inch sheet iron pipe for the Spring Valley Water Works. Its crowning achievement was the Virginia City, Nevada, water works. There, Risdon installed all the hydraulic machinery: pumps, engines, accumulator (all capable of working under a pressure of 2,000 pounds per square inch), as well as 36,000 feet of sheet iron pipe. A similar system was also installed at Central City, Colorado. So advanced was Risdon's design of iron pipe, that its manufacturing techniques were employed for pipes going into an aqueduct for Washington, D.C.[53]

By 1872 the company had changed its name to the Risdon Iron and Locomotive Works, and its location to the corner of Howard and Beale streets. With a work force of skilled foundrymen and mechanics, manning the most up-to-date equipment, it had the capacity to build just about anything, and did. Car wheels were its principal product for railroads, the client list of which was impressive: the Central Pacific Railroad, the Southern Pacific Railroad, the Northern Pacific Railroad, as well as numerous city railways. In 1872 it constructed two vessels: a 500-ton steamer for transporting coal from Oregon, and a 235-ton revenue steamer.

Aside from doing work for shipping companies, the Comstock mines were Risdon's principal customers. In 1871 it constructed two of the largest quartz mills in Nevada: the Brunswick and Union mills, both 60-stamp installations. And the Belcher Mine took two 20 x 40-inch pumping engines with boilers, and two 18 x 27-inch engines for deep hoisting.[54] The following year Risdon constructed three more hoisting engines for the Comstock: a 20 x 30-inch engine for the Gould & Curry Mine, and two for the Sutro Tunnel, each capable of hoisting from a depth of 2,000 feet. At the same time, Risdon shipped a 12 x 24-inch hoisting engine for the Phoenix, Erie and Original Amador Mine, as well as a large quartz mill for the Floral Mill at Pioche, Nevada. An engine capable of pumping from a depth of 3,000 feet was fabricated for the Imperial Mine at Washoe.[55] In all, Risdon sold $200,000 worth of mining machinery to Comstock companies in 1872.[56]

Risdon's work force in 1874 reflected the lucrativeness of the mining machinery business. Three hundred men worked ten hours a day, six days a week, fabricating equipment mainly for Comstock mines

"The annexed excellent engraving represents the well-known and one of the earliest manufacturing establishments on this coast. Messrs. Coffey and Risdon, the enterprising proprietors, have for the past ten years devoted their entire attention to the manufacture of steam boilers and sheet iron work. Their efforts have been crowned with success and the Government has frequently entrusted them with work in their line. It appears also that the contracts and work they have undertaken have been most satisfactorily executed. Messrs. Coffey & Risdon have added considerably to their establishment. Adjoining the workshop, which fronts on Market, Bush and Front streets, they have secured an extension lot for their iron yard, and now the premises occupy an area of 160 feet front by 137 deep. They give employment to 120 experienced workmen, and from early dawn to late eve, the busy hum of the hammer seems ceaseless. They have just finished a large order for boiler work, which is now being shipped to the Honolulu Sugar Refining Company. It comprises a series or number of boilers, tanks, vacuum pans and heaters. Large works of a similar nature were recently shipped to the Honolulu Tallow Refinery; all of this work is made in the most approved and substantial manner. They have constantly on hand from eight to ten boilers, some of which are for the California Steam Navigation Company. Many orders now reach them from all parts of this coast for boilers for various purposes, which certainly proves that their work stands in high repute for strength and durability. Besides this, they keep for sale, and have perhaps the only complete assortment of sheet iron, iron pipe, rivets, etc., on the Pacific Coast, which they will dispose of at reasonable prices."

This depiction of Coffey & Risdon's Iron Works, with accompanying write-up, appeared in the *Mining and Scientific Press* on July 2, 1862.

and mills.[57] Fifty additional mechanics were added the following year to construct hydraulic equipment: a 600 horsepower compound pumping engine for the Belcher and Crown Point mines, capable of lifting 750 gallons per minute from a depth of 3,000 feet.[58] That was a tiny Cornish engine compared to the beam compound direct-acting pump designed by W. H. Patton for the Imperial Mine. That one expelled 60,000 gallons of water per hour.[59] Risdon also built a similar pumping outfit for the Lady Bryan Mine, which was preparing to sink 2,000 feet.[60]

While other San Francisco foundries languished during 1878–79, a slack period caused by the near collapse of mining on the Comstock, Risdon managed to keep busy. In 125 days it put together for the Yellow Jacket Mine one of the largest hoists every assembled on the West Coast. It consisted of a pair of direct-action Corliss engines with reels, brakes, and associated paraphernalia. Designed by W. H. Patton, the whole outfit weighed over 350 tons. In 1882 Risdon put together another pumping outfit for the C. & C. combination shaft on the Comstock. This machinery was placed on the 2,400 foot level.[61] A year later another Risdon pump, capable of raising 2,000 gallons per minute in a single lift of 1,030 feet, was installed in the Sutro Tunnel.[62] During this period, Risdon filled numerous small orders, mostly hoisting engines of the compound variety, which had nine- and eighteen-inch cylinders. These engines were compact, one piece units, capable of working from 1,200 feet with great efficiency.[63]

In common with other foundries, Risdon produced a line of small, economical mining appliances designed for prospecting and small mine application. Such was

At left: The all-steel "Common-sense" Horse Whim. The arms of the wheel were bolted to the hub and to the rim of the drum, which allowed for easy disassembly for transportation on pack-animals. The one-horse whim had a six feet diameter drum winding 500 feet of rope. The sweep could be thrown out of gear by forcing a wedge-shaped piece of iron under the shaft. If the horse would not stop upon getting the bucket up to the sheave, the mechanism could be thrown out of gear, allowing the horse to go around in circles, the automatic grip holding the load. With an ordinary-gaited horse, the speed of the hoist, was eighty feet per minute, or about 100 buckets of ore every ten hours.

the "Common-sense Horse Whim." This all-steel portable hoist weighed no more than 650 pounds. It was sectional, no piece weighing over 100 pounds, and could be packed anywhere a jack could go. Costing $150, it was touted as being able to hoist "an oil barrel filled with horseshoes from a depth of 300 feet."[64] Risdon also made small, sectional water-jacket silver-lead smelting plants similar to those produced by the Pacific Iron Works. These came equipped with Baker blowers, and an improved Corliss engine and boiler.[65]

In the late 1880s Risdon introduced the Bryan Roller Quartz Mill, an appliance intended to replace the traditional gravity stamp mill. A modification of the well-known Chilean mill, it consisted of three thirty-inch steel-tired rollers, each weighing 1,200 pounds, running over steel dies in an annular mortar. One set of tires would crush 1,600 tons before replacement. The mill ran at fifty revolutions per minute and was said to be almost noiseless. It was perfect, so the advertising claimed, for continuous wet-crushing of gold or silver ores. The unit was portable, could be quickly and cheaply erected, and required less power and initial startup costs than any stamp mill. Requiring only five horsepower to operate, the mill could crush fourteen tons a day with a No. 40 screen. This roller mill could be adopted to any size gold mine, large or small.[66]

Embraced by a great many mining companies, this mill became a Risdon specialty. In March 1889 the company sold two dozen of them and had orders for fifteen more.[67] For several years Risdon shipped mills throughout the Western territories and states, as well as to Virginia, Georgia, Mexico, Tasmania, and Australia.[68] It was apparent by March 1893 that there was "quite a little boom in Bryan Quartz mills."[69] That

Below: The Bryan Roller Quartz Mill, manufactured by the Risdon Iron Works, was simple in construction, and required no framework, being built ready for installation upon its foundation. In the operation of this mill, the pulp flowed around the mortar next to the screens in a rapid current of not less than 300 feet per minute; but toward the center, inside of the rollers, the movement of pulp was much slower, and as gold was liberated from the matrix it fell to the eddy side of the current, and was amalgamated around the cone in the center of the mortar. It is said that the movement of the mill was almost noiseless at fifty revolutions per minute. The mill was economical, requiring less than five horsepower to run, and would crush about fourteen tons a day through No. 40 screens. Capacity of the mill could be varied by adding weight to the tank pulley, which rested on top of the rollers. (*Mining and Scientific Press*, June 11, 1887)

When carbonates of lead or copper ores contained a sufficient percentage of metal to smelt without addition of flux, their treatment in water-jacket furnaces was the simplest form of smelting, requiring nothing but fuel and ore. Frequently the percentage of metal was too low to smelt, or the ores contained copper pyrites, glance, and galena. Such occurrence thrust miners into the complex side of the smelting process where mixtures of iron ore, limestone or silica were necessary. The correct percentage of these fluxes added to the ores had to be determined. This problem was sometimes simplified if ores were present that fluxed each other.

To the left is a galena silver-smelting furnace, with a water-jacket rectangular in form, built in sections and capable of being extended, so as to give a daily capacity of sixty tons of ordinary smelting ore. Manufactured by the Risdon Iron Works, these smelters were adopted by the best Eastern reduction works, and gave excellent results. The regular sizes were thirty-six by forty-two inches, up to thirty-six by seventy-eight inches, although special sizes were built on request. (*Mining and Scientific Press*, January 7, 1893)

spring the company began distributing the Johnson concentrator.[70] And it still turned out all styles of mining hoists powered by Corliss engines.[71] The real feather in Risdon's cap at this time was the building of the world's largest mill.

In the early 1890s the Alaska-Treadwell Gold Mine placed an order with Risdon for a mammoth mill consisting of 240-stamps. Fabrication and erection of this mill took several years, and prolonged Risdon's mining equipment trade beyond that of other San Francisco suppliers. It allowed Risdon time to alter its operation to meet the threat of competition.

Aware that Eastern machinery manufacturers were cutting into markets that San Francisco companies had long considered their territory, Risdon chose not to fight the competition. Instead, it joined it. In the 1890s it began carrying a line of Eastern components such as engines, pumps, blowers, gas engines, and water gates. All manner of small things were stocked and drawn upon as need arose, to fill orders for min-

To the right is a high-class Cornish roller manufactured by the Risdon Iron Works. Cornish rolls were sometimes used instead of stamps for dry pulverizing of ore after it had passed through a rock breaker. Two sets were generally used — one for roughing and one for finishing. Roughing rolls took ore from the crusher, delivering it through revolving screens, from whence all that is too coarse for the roaster or jig was returned to the finishing rolls, which were adjusted to reduce the ore to the desired fineness. Rolls were used to best advantage in large mills, where they worked continuously. Risdon made ten sets of these rolls for the Anaconda Mine, at Butte, Montana.

Another type of improved ore crusher fabricated by the Risdon Iron Works, similar in design to those manufactured by the Fulton Iron Works. (Both illustrations on this page are from the April 25, 1891, issue of the *Mining and Scientific Press*)

ing, milling, and other machinery — orders which Risdon would have lost to Eastern contractors.[72]

Sensing the demise of San Francisco's mining equipment trade, Risdon sought to supply the needs of emerging electric railroads by fabricating engines, condensers, switches, and overhead lines and poles. It was successful in this line, furnishing 500 and 1,000 horsepower engines to the San Francisco and San Mateo electric road, the Metropolitan electric road, the Oakland, Hayward & San Leandro road, Santa Cruz road, and the Stockton road. Diversification didn't stop there.

Risdon contemplated the construction of ocean-going vessels, if only it could find a suitable location for plant and drydock. Several municipalities offered land. The city of Vallejo, on the straits at the head of San Pablo Bay, opposite the Mare Island Navy yard, tendered forty acres, all the necessary water frontage, and a handsome bonus as well. Oakland feebly promised aid and its Board of Trade passed a series of reso-

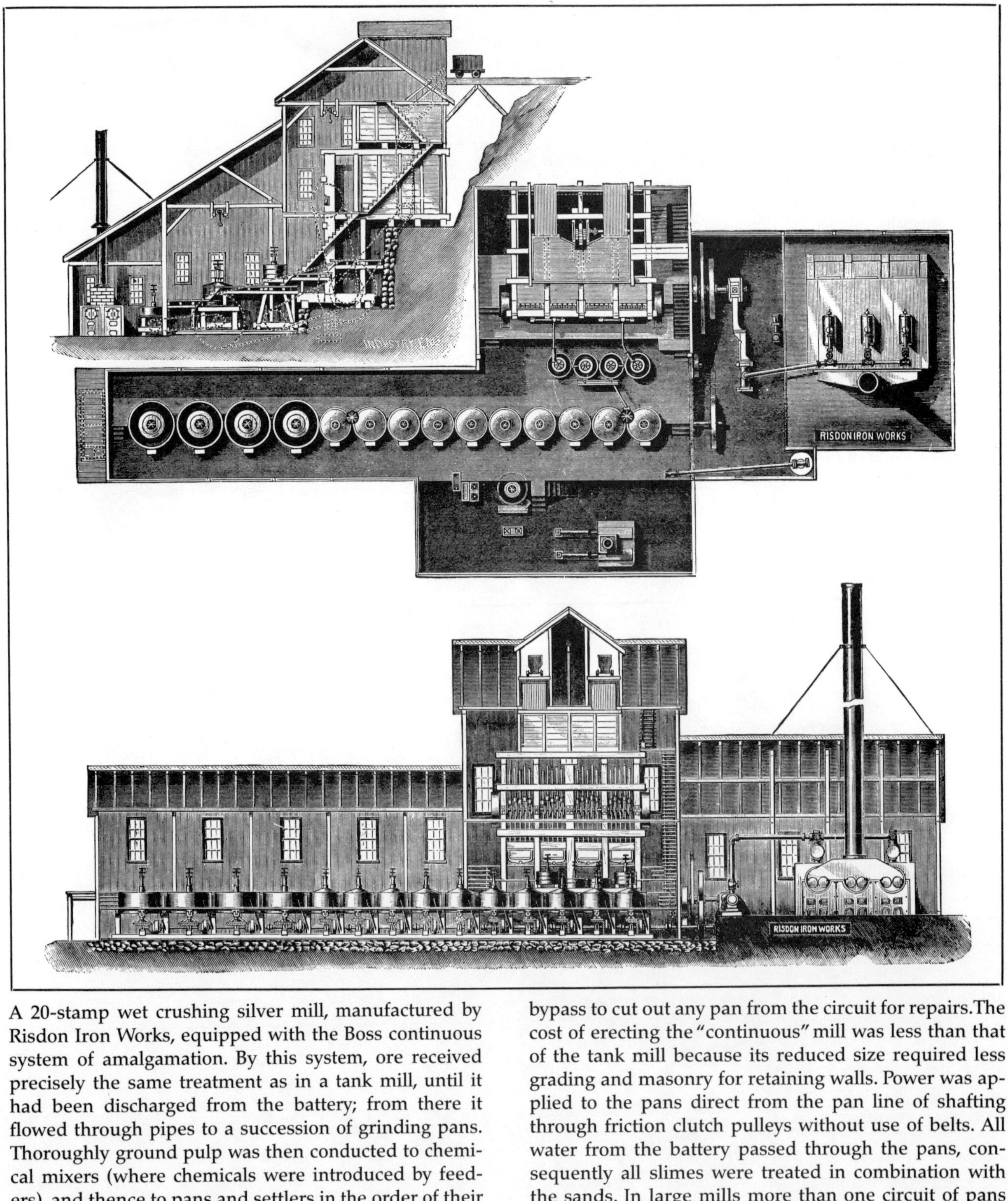

A 20-stamp wet crushing silver mill, manufactured by Risdon Iron Works, equipped with the Boss continuous system of amalgamation. By this system, ore received precisely the same treatment as in a tank mill, until it had been discharged from the battery; from there it flowed through pipes to a succession of grinding pans. Thoroughly ground pulp was then conducted to chemical mixers (where chemicals were introduced by feeders), and thence to pans and settlers in the order of their arrangement. Distribution pipes introduced quicksilver into either of the amalgamating pans as desired. Steam syphons were provided to clean the pans, and for a bypass to cut out any pan from the circuit for repairs. The cost of erecting the "continuous" mill was less than that of the tank mill because its reduced size required less grading and masonry for retaining walls. Power was applied to the pans direct from the pan line of shafting through friction clutch pulleys without use of belts. All water from the battery passed through the pans, consequently all slimes were treated in combination with the sands. In large mills more than one circuit of pans and settlers were employed, each circuit treating a portion of the pulp separately. (*Mining and Scientific Press*, April 9, 1892)

A cross section of the world's largest quartz mill. Built by Risdon Iron Works for the Alaska Mining & Milling Company at Douglas Island, Alaska, the mill originally had 120 stamps, each stamp weighing 850 pounds. The number of stamps was subsequently doubled. Equipped with six rock breakers and ninety-six concentrators, the mill processed 600 tons of ore daily. The mill was driven by water wheels during the wet season, and by steam engines during dry periods. (*Mining and Scientific Press*, September 20, 1890)

lutions. The decision to enter shipbuilding was made, and in fall of 1900 Risdon purchased the property of the old Pacific Rolling Mill in Potrero, consisting of thirty-two acres. Adjoining the Union Iron Works, this land had 1,700 feet of deep water frontage. At the same time, application was made to the Board of State Harbor Commissioners for a drydock site. While this was going on, W. H. Taylor, Risdon's president, and L. R. Meade, secretary, were in Washington, D.C., submitting estimates for two cruisers.[73] The Vallejo drydock site was approved, and in January 1901 the company began constructing a first class shipbuilding facility with wharves, machine shops, and overhead cranes.

With two shipbuilding sites — one on the Potrero, the other at Vallejo — Risdon rapidly caught up with the Union Iron Works in the race to become premier shipyards. And in so doing, both companies became targets for corporate takeover. In 1905 Bethlehem Steel Company absorbed the Union Iron Works, and six years later Risdon fell prey to the giant Eastern company.[74]

Regardless of takeover, Risdon maintained its name until after World War I, and like the Union Works, continued producing mining and milling machinery. For three decades following its plunge into shipbuilding, Risdon marketed the Johnston concentrator, as well as air compressors, hydraulic plants, boilers and engines. Risdon's electric hoists revolutionized ore haulage on the Comstock, and its giant dredges sucked gold-bearing gravels in California, Montana, Alaska, Africa, and Central and South America.

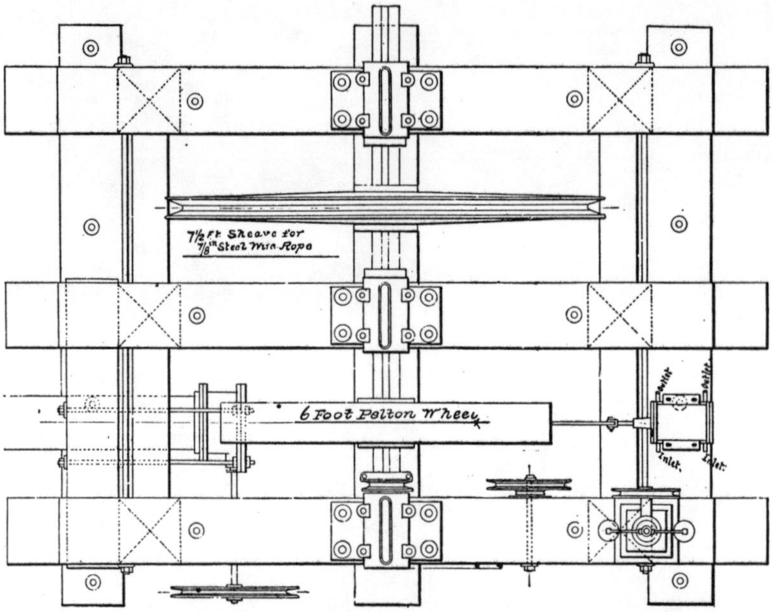

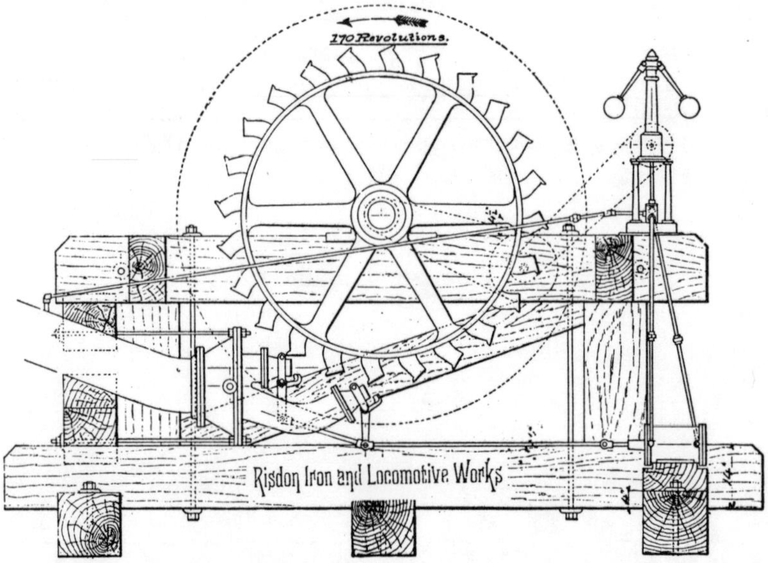

The use of water wheels greatly reduced the expense of milling in California, and in some places water was brought a considerable distance in pipes or flumes, at great expense. The favorite wheel used in California was that known as the hurdy-gurdy, a simple and efficient appliance, especially adapted to high heads of water pressure. Two forms of hurdy-gurdies were used: the Knight and the Pelton, the latter used in most quartz milling. The Pelton wheel was used throughout California's Mother Lode, as well as in South Africa, Valparaiso, New Zealand, and Alaska. The illustrations at right show a Pelton wheel, with automatic governor, adapted to run a 60-stamp quartz mill built by the Risdon Iron Works. Such wheels were also utilized to drive air compressors, pumps, sawmills, hoisting machinery, factories, mills, and printing plants. (*Mining and Scientific Press*, July 11, 1885)

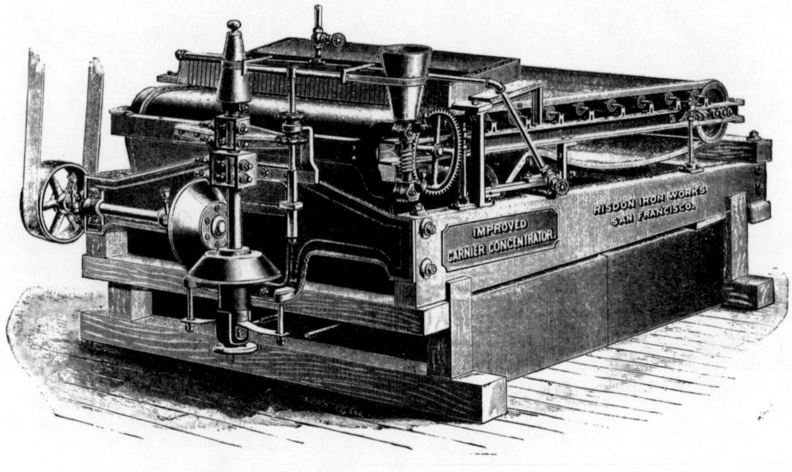

At left: The Garnier Ore Concentrator, made by the Risdon Iron Works, was said to have a circular movement corresponding to that of a "batea." The vertical shaft seen at front of the machine, had an eccentric bearing at the top, which gave a rotary movement to the table. Pivot studs supported the rear of the table, and a sliding bearing in the center permitted a forward and backward movement equal to the throw of the eccentric in front, but no movement sideways. *Mining and Scientific Press*, January 17, 1891)

CHAPTER FOUR

Aetna and Vulcan Foundries, Parke and Lacy, the Pelton Water Wheel Company, and the Joshua Hendy Iron Works

THE Aetna Iron Works was founded in early May 1866 by four master mechanics who had worked in other San Francisco foundries.[1] Wallace and John Hanscom arrived in California from Connecticut about a decade earlier. They combined their talents and resources with those of thirty-four year old Henry S. Smith, a machinist from Rutland, Vermont, who had worked in the Vulcan Iron Works, the Pacific Iron Works, and at the Golden State Foundry.[2] The fourth man, James Pendergast, was a newcomer to the foundry scene, having arrived in San Francisco in 1864. He had previously worked for the Union Iron Works.[3]

The firm erected several large buildings on the corner of Tehama and Fremont streets. Its two-story machine shop fronted on Fremont, and the forty by fifty foot foundry housed a McKensie blower and a cupola having a capacity of six tons of pig iron a day. While they were well prepared to turn out light or heavy machinery "in the finest of manner," quartz mills was their specialty from the start.[4]

Aetna Foundry introduced what it termed "a new" quartz mill in December 1867, consisting of 10-stamps complete with engine and other machinery. The stamps weighed 650 pounds each and worked in two copper-lined mortars. The engine had a twelve-inch cylinder and a twenty-four-inch stroke. Steam was provided by a boiler that was sixteen feet long by forty-eight inches in diameter. It had forty-six tubes. The drive belt, linking engine to mill, made by M. M. Cook & Son, was of the best oak-tanned California leather. Every piece of leather going into the belt was carefully selected to provide uniform wear and tension. (A leather belt, it was claimed, when properly made and of the right material, was superior to rubber which was utilized for most belts than driving heavy equipment.) The only novel thing about the mill was that its stamps could be held up for repair by passing an iron shaft across the back of the battery to engage the tappets. Thus any, and all stamps, could be securely held up for repairs. It was an innovation soon adopted by all manufacturers of stamp mills. The mill was fifty feet long by ten inches wide, and copper fastened. The first mill of "new design" was shipped to the Merrimac Mining Company, fourteen miles southeast of Oroville in Butte County, California.[5]

In 1869 manufacture of Varney pans was transferred from the Miner's Foundry to Aetna, and a year later the company broadened its machinery line by introducing the Varney and Rix rock crushers, engines, hoists, and Pendergast's superior white iron shoes and dies for stamp mills.[6] Interestingly, all of this machinery was photographed as fast as it was produced. Although photographing equipment was a common Eastern practice, it had never been done on the West Coast; another first for Aetna.[7]

The White Rotary Pulverizer, an invention of George W. White of San Francisco, was manufactured by the Aetna Iron Works. Tested in early 1880 at the Alexander Mining Company, Grantsville, Nevada, this machine was supposed to overcome many of the defects inherent in rotary pulverizers, such as lack of free discharge, heating, friction, and wear due to high speed. White overcame these problems with a new type of friction gear (shown in engraving), and by allowing passage of highly pulverized ore not through wire screens or punched plates, but through the beater bars.

The bed plate of the pulverizer was cast in three pieces, bolted together. Chairs for the friction rollers were planed on the bottom, and their seats on the bed plates were also planed. The machine was entirely open on the bottom of the cage, the center piece of the bed plate being cut out five inches wider than the cage which carried the beater bars.

Ore attrition was accomplished by five rows of beater bars, made of the best homogeneous steel, rotating within the cage, three rows running in one direction and two alternate rows running in the opposite direction. Ore for the pulverizer was first passed through a rock breaker to reduce it to a size suitable for the screw feed attached to the pulverizer. This feeder distributed the ore in equal quantity to each compartment of the cage. It delivered pulp of a uniform size, suitable for concentration, roasting, or lixiviation.

The machine was fabricated in a fifty-six inch diameter model, and was available in sections so small as to be easily packed on animals. (*Mining and Scientific Press*, March 20, 1880)

The company also carried on considerable architectural work, casting iron columns for the Jewish Synagogue erected on Taylor Street, and fabricating the iron work for the Roman Catholic Orphanage at Bay View. A new departure for Aetna, however, was the manufacture of the Hyde Road Steamer, a traction engine designed for farm work and transportation of cargo over rough terrain. The steam pump for the engine, invented by the brothers Hanscom in June 1871, was an adaptation of a previous automatic steam vacuum pump, claimed to be no more than a hollow cylinder having no piston or other mechanism.[8]

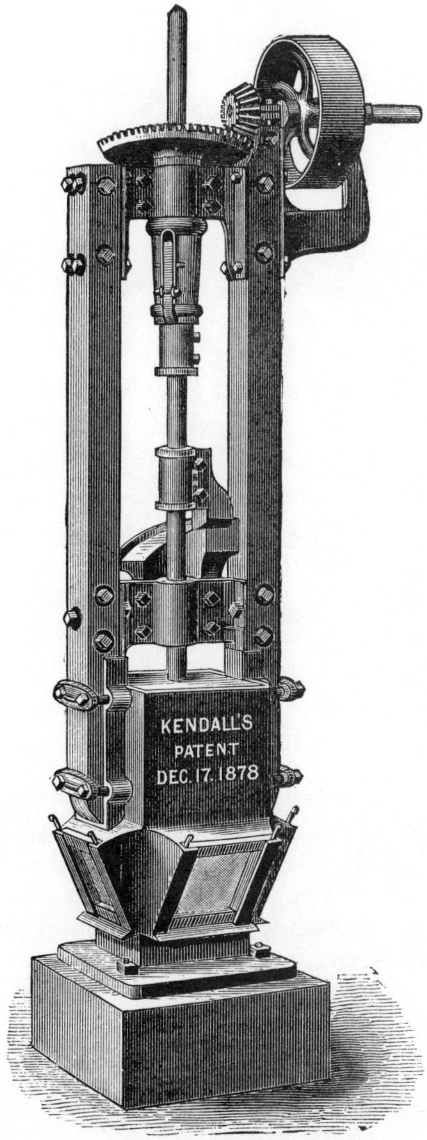

The Kendall One-stamp Mill, produced by the Aetna Iron Works, was considered a radical design, in that it had a novel curved incline cam mounted on a rubber cushion to prevent jarring, and a peculiarly shaped tappet. The latter was secured to the stem of the stamp, so as to revolve as it was lifted, which gave the stamp both a crushing and a grinding action.

Constructed for one stamp only, the mortar had three discharge openings fitted with screens. The Kendall Mill was fabricated complete and ready for instant shipment, and came in one and three tons capacity models, made in sections for mule packing. The mill was intended for prospecting, but additional stamps could be added, one stamp at a time, as the mine developed.

The Kendall Mill seems to have been accepted. In February 1882 there were three in use in Arizona, two in Nevada County, California, three in Amador County, two in Tuolumne County, and several in San Francisco. Others were sent to Montana, Idaho, British Columbia, and Honduras. (*Mining and Scientific Press*, February 11, 1882)

This horizontal flume was patented by A. Myers in 1873, and manufactured by the Aetna Iron Works. Coupled with the Leffel Water Wheel, the flume made for a powerful device to drive quartz and saw mills. It consisted of two semi-spheres flattened at their sides, giving a large water passage opposite the buckets of a turbine, or where water was admitted to the wheel. Coupled with a fifty-two inch Leffel Wheel, the Myers Flume was first used at the Eureka Mill on the Carson River, where it developed power equivalent to 444 horsepower. It was next used at the Empire Mill at Plymouth, Amador County, California, where a Leffel Wheel drove an 80-stamp mill. (*Mining and Scientific Press*, May 15, 1880)

Apparently the Hanscoms put their expertise to work building steam-powered stamps of the Wilson design, which were first produced by the Pacific Iron Works. These small two-stamp mills were suitable for prospecting and small mine development. Such a unit was shipped to Globe, Arizona, in May 1877.[9] In 1880 the company introduced a large rotary pulverizer invented by George W. White of San Francisco.[10] Two years later, it produced the Kendall Quartz Mill, a steam-powered, one-stamp unit also designed for prospecting. It was sectional and could be packed anywhere.[11]

Although Aetna did a great deal of general foundry work, and its quartz mills were highly regarded, the company faltered during the hard times following collapse of silver mining in the mid-1880s. Orders slacked and the company did not survive into the 1890s.[12]

The firm of Parke and Lacy began modestly about 1874, not as a foundry and manufacturing company, but as representatives for Burleigh drills and air compressors, the Putnam Machine Company's machinists' tools, Haskins engines, Wright's steam pumps, and miscellaneous mining machinery. Well versed in mining technology, its founders had an impact on Western mining long before setting up shop in the Bay City.

The oldest member of the firm, Lyman Curtis Parke, was born at Pontiac, Michigan, in 1834, one of five children, four of whom became prominent miners. Three Parke brothers — Frank, Ira, and Lyman — were well known on the Comstock. And Lyman, along with brother Harvey, had experience in the Lake Superior copper region.

In 1870 L. C. Parke associated himself with John Waring in New York. Their firm, Waring-Parke Company, manufactured the Waring air compressor, a device later taken over by A. C. Rand of the Rand Drill Company. It was sales and service of air compressors that brought Parke to the Comstock and led to his association with Benjamin Thomas Lacy.[13]

New Englander Benjamin Lacy was born in 1845. A machinist by trade, he helped Charles Burleigh perfect his pneumatic drill and air compressor. Lacy was in charge of the Burleigh drill crews working at the Hoosac Tunnel in Massachusetts, and subsequently assisted in the removal of the Hell Gate Reef in New York Harbor. He demonstrated Burleigh drills at the Mont Cenis Tunnel and put those machines to work in Comstock mines and the Sutro Tunnel. The firm of Parke and Lacy was Comstock bred in 1873.

Below: James Leffel's Vertical Mining Water Wheel, was manufactured by Aetna Iron Works and distributed by Parke and Lacy. Placement of pulley on the wheel's horizontal shaft afforded the most efficient means for transmittal of power. Properly proportioned pulleys and light but sufficient belting controlled the amount of torque. The wheel was applicable not only to mining but to the driving of sawmills and other machinery. (*Mining and Scientific Press*, October 1, 1881)

Parke and Lacy established their mining machinery business in San Francisco about 1874. They started on a small scale as agents for Burleigh drills and compressors, Haskins engines, Wright's steam pumps, and Putnam machine tools. In two years they built an extensive trade, which forced a move to larger quarters. In February of 1876 the firm moved into this building at 417 Market Steet, occupying two floors. Parke and Lacy displayed their equipment on the main floor, and kept compressors and heavy machinery in a nearby warehouse. Ten carloads of Burleigh drills were kept on hand at all times. The building was also shared by the West Coast offices of Fairbanks and Hutchinson Scale Company. This company, whose manufacturing plant was in St. Johnsbury, Vermont, produced all sorts of weighing devices from small assay balances to large scales capable of weighing railroad cars. Fairbanks & Hutchinson controlled the scale trade on the Pacific Coast, in China, Japan, and the Pacific Islands. The F. Ogden Company, representatives for Tulloch Ore Feeders, also occupied a portion of this building. (*Mining and Scientific Press*, February 5, 1876)

As Burleigh's West Coast representatives, Parke and Lacy introduced rock drilling machinery to California.[14] They opened a San Francisco office in the Hayward Building at 6 Leidesdorff Street. Two years later they moved to 417 Market Street, where they opened a "commodious salesroom," displaying their equipment.[15]

Priding thesemselves on keeping abreast of mining technology, Parke and Lacy introduced in 1877 the Walker compound steam pump, designed for heavy pumping of dirty water.[16] The following year Parke toured Europe to study continental rock drilling equipment. He found "nothing that American mining wanted." European rock drilling machines, he reported, were "clumsy and cumbersome in the extreme."[17]

Parke and Lacy had no manufacturing adjunct. They were machinery dealers only, and as such, jobbed out orders that required special construction. One such order was for a complete mill, with engine, roasting furnaces, and lixiviating works, for the Rosario Mines

Lyman C. Parke, a native of Pontiac, Michigan, was a successful mechanical and mining engineer, a practical miner, and an originator and manufacturer of mining equipment. He died in Oakland, California, on November 11, 1899. (*Mining and Scientific Press*, November 18, 1899)

on a spur of the Sierra Madre, 100 miles from the port of Ajiavampo, on the Gulf of California. That order was sublet to the Union Iron Works in late 1880. This $150,000 order was put together at the foundry, tested in San Francisco, then disassembled and loaded aboard a small fleet of schooners and shipped to Guaymas, a port of entry. From there, the equipment was transported 150 miles down the coast to Ajiavampo, where it was unloaded. Because the mill had to be hauled inland a hundred miles, Parke and Lacy had made and shipped aboard the steamer *Newbern,* ten sixteen-mule wagons with harness and all necessary appliances for handling the machinery. Special wagons with saddles were made to accommodate the 11,000 pound engine frame, and two steel boilers which weighed 7,500 pounds each.[18] Such orders obviously required considerable personal supervision on the part of the contractors.

Parke and Lacy also sold the Huson aerial tramway, a number of which were in use in Colorado by 1887.[19] And they were agents for the Eastern manufactured Standard single- and double-geared hoists which were designed for small and medium size operations.[20] For deep mining Parke and Lacy offered the Lidgerwood hoisting engines, manufactured in Brooklyn, New York.[21] A variety of mills and crushers were also offered, including 2- and 3-stamp units for prospecting and small mine operation.[22]

Parke and Lacy's mainstay, however, was the Dodge Improved Pulverizer. Invented by Miles B. Dodge and manufactured in San Francisco at the Empire Iron Works, this machine was no more than a hexagonal steel barrel lined with steel bars, which revolved at thirty-two revolutions per minute. It could be used for either wet or dry crushing.[23] The most unusual crusher in their line was the "Cyclone Crusher," or Wall Crushing Rolls. No more than a Cornish roller with fluted grinding surfaces, this device was designed to crush sandstone containing silver chlorides at Silver Reef, Utah. Patented in 1886–87 by E. A. Wall, it found wide application throughout Utah, being used at Milford, Park City, and at sampling works in Salt Lake City. Several were sent to New South Wales, and to Maryborough, Queensland, and to Adelaide, South Australia.[24]

Parke and Lacy grew rapidly, opening outlets in

By 1876 Parke and Lacy had introduced 150 Burleigh drills and compressors to West Coast mining. More would follow. The Burleigh consisted of a compressor, drill, and various forms of carriages or holders upon which the drills were mounted. The drill shown at left is mounted on a mining carriage, which consisted of an upright column with a screw clamp nut for holding, raising, and lowering the drill; the whole mounted on a platform, which slid by means of a crank across the carriage, allowing adjustment of the drill to any position on the heading. It was secured in position by the jack-screw at top of the column and could be easily moved to permit blasting.

The Burleigh mining drill produced bores from one-and-a-quarter- to two-inches, and fed twenty-six inches without change of drill points. The drill weighed 475 pounds and required only two and a half horsepower to run. This style of Burleigh drill was popular in California mines between 1874 and 1876. It was used in four to five foot adits, and eight by ten foot drifts.

Below: A style of Burleigh drill carriage, known as a "Buffalo," introduced by Parke and Lacy about 1874. This carriage mounted two drills upon one bar, which could be raised or lowered by means of chain pulleys and a windlass. It was constructed with an open space of from four to sixteen feet in the clear, between the two sides, so that it could be run in as soon after a blast as track could be cleared by throwing the rock into the center. This style of carriage was used at the Sutro Tunnel, the Belcher, Gould & Curry, Consolidated Virginia and the Overman mines on the Comstock.

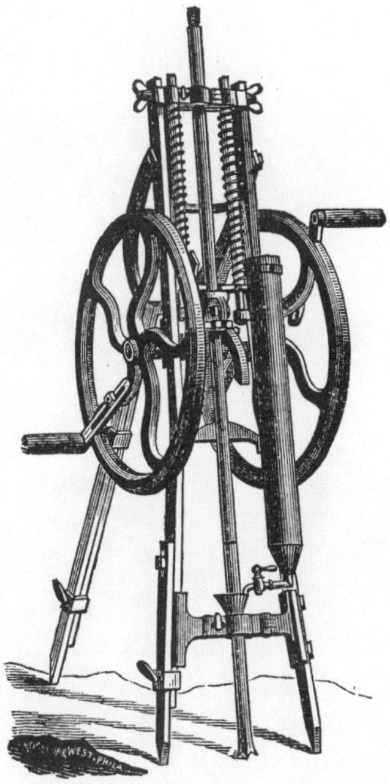

Parke and Lacy also represented the Victor Rock Drill, above. A machine of simple construction, this drill could be run either by hand or steam. Weighing only 350 pounds, it was readily moved by one person in less than a minute, by laying it down and running it upon its own wheels, like a cart, from place to place in a mine or quarry. The machine was self-feeding and self-cleaning. (*Mining and Scientific Press*, August 27, 1881)

By the early 1880s Parke and Lacy represented nearly all manufacturers of rock drills. To represent Ingersoll Rock Drills, the company acquired the Berry & Place Machine Company, at 8 California Street in San Francisco. The Ingersoll Drill soon outstripped the Burleigh in mine application, mainly because of its fewer moving parts, which resulted to less repairs and expense. Shown at left is an Ingersoll rock drill mounted on a single carriage. At right is another view of the Ingersoll drill. Above: The Ingersoll-Sergeant air compressor, also a Parke and Lacy item, drove pneumatic drills and provided compressed air for other mine applications. This was a duplex Corliss air compressor, with compound steam cylinders and piston inlet air cylinders. A condenser was used where water supply permitted. (*Mining and Scientific Press*, August 19, 1882, and June 11, 1892)

Besides mining equipment, Parke and Lacy sold the "Plantation" Circular Saw Mill. Built by Dennis Lane Manufacturing Company, this mill would accommodate saw blades up to fifty-four inches in diameter. With wood frame, twenty-five foot carriage, but minus the blade, these mills sold for $350. Although designed to meet the needs of small Southern lumbermen, as the name implied, the unit was used throughout the Far West, particularly for sawing mine timbers. (*Mining and Scientific Press*, January 21, 1882)

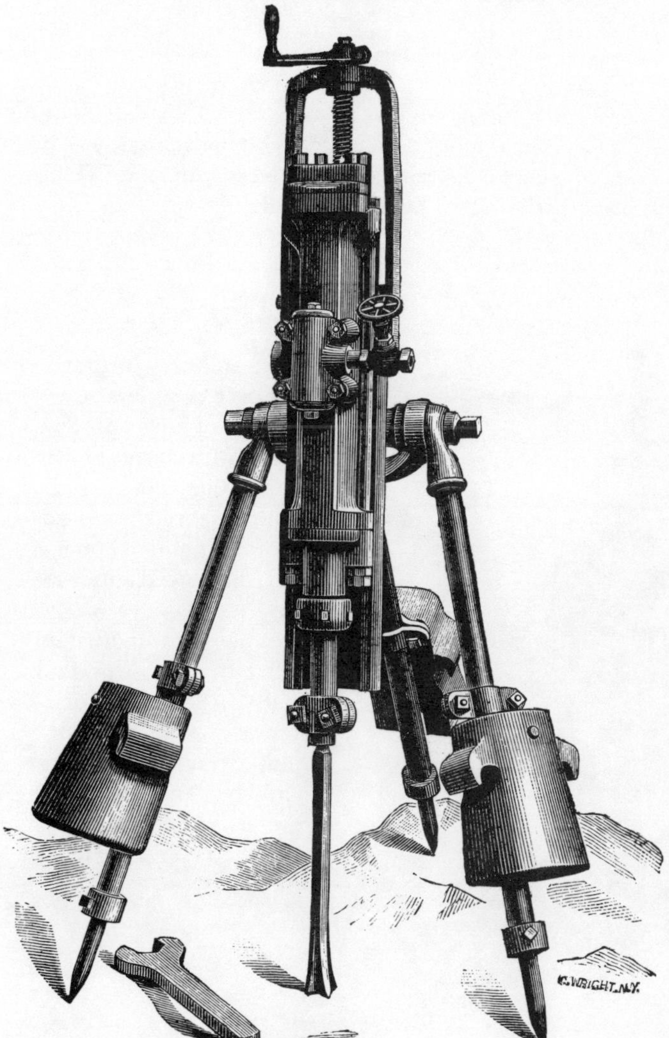

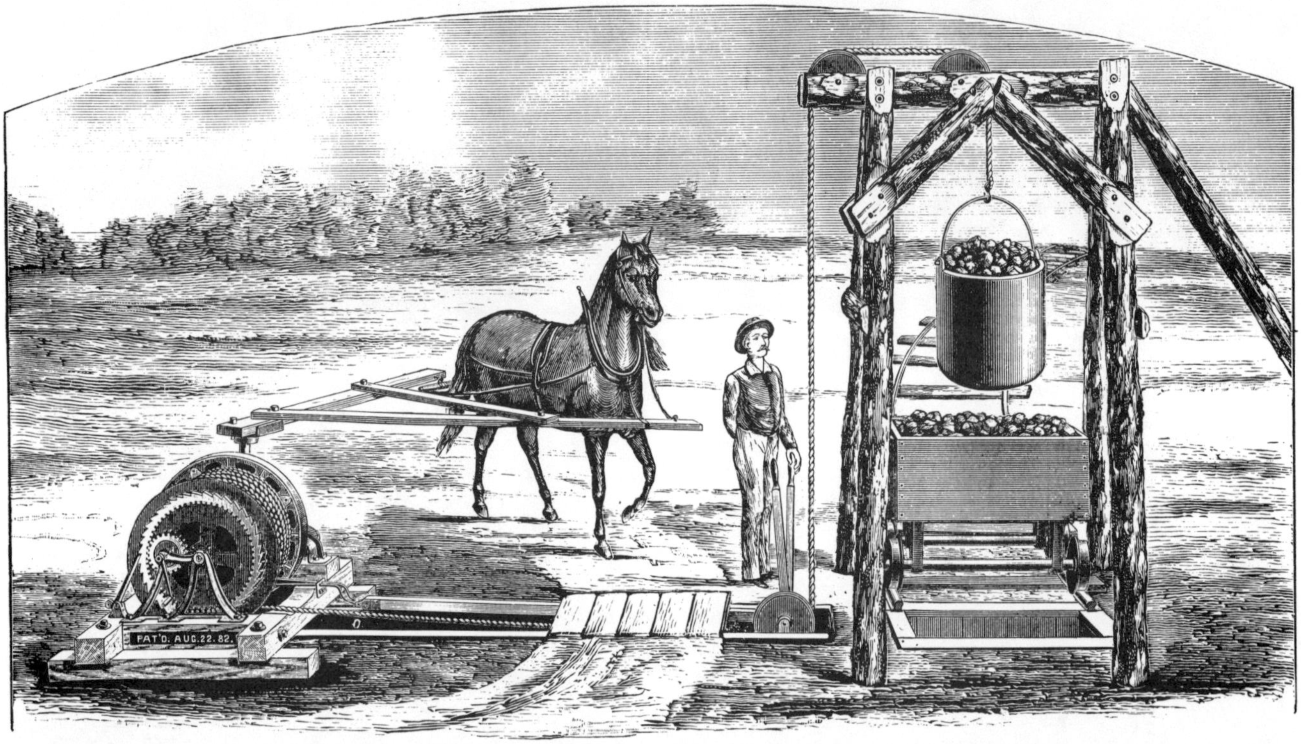

Parke and Lacy also sold this horse powered hoist. It would raise a bucket, or a load weighing from 500 to 700 pounds, sixty to seventy feet per minute. Made entirely of iron and steel except the sills, the machine was advertised as not being affected by dry or wet climates. The hoisting drum was controlled by a man at the shaft, who landed the bucket by operating the levers at hand, by which means the drum was thrown in and out of gear at will while the horse was in motion. The bucket could be lowered by the brakes as safely and conveniently as by a steam powered hoist, so the advertising said. (*Mining and Scientific Press,* January 19, 1884)

Parke and Lacy introduced to West Coast mining a number of hoists produced in the Midwest and East, among them single and double-geared Standard Mining Engines shown at left. Introduced about 1887, these outfits were adapted to mines of ordinary depth, with either single or double shafts, or to inclines or tramways with single or double tracks. The engines could be altered so that the operator could stand at either side or in the rear of the drum.

The Niles double-geared hoists consisted of a pair of Corliss engines, bolted at each end of a cast-iron sole plate. Parke and Lacy furnished these hoists in the following sizes: 8 x 12-inch engines, drums 48 x 28-inches, weight 15,000 pounds; 10 x 12-inch engines, drums 48 x 28-inches, weight 18,000 pounds; and 12 x 18-inch engines, drums 66 x 30-inches, weight 40,000 pounds. (*Mining and Scientific Press*, July 24, 1880)

The Standard Mining Engine were double cylinders, with double or single drums, as required. The drums were either straight or cone-shaped, and smooth or grooved. On power plants of thirty horsepower and over, the makers usually put in double gearing, as shown at right, with drums keyed to the shaft, and foot brakes on both crank wheels. Extra length or extra diameter drums were furnished if desired. (*Mining and Scientific Press*, December 31, 1887)

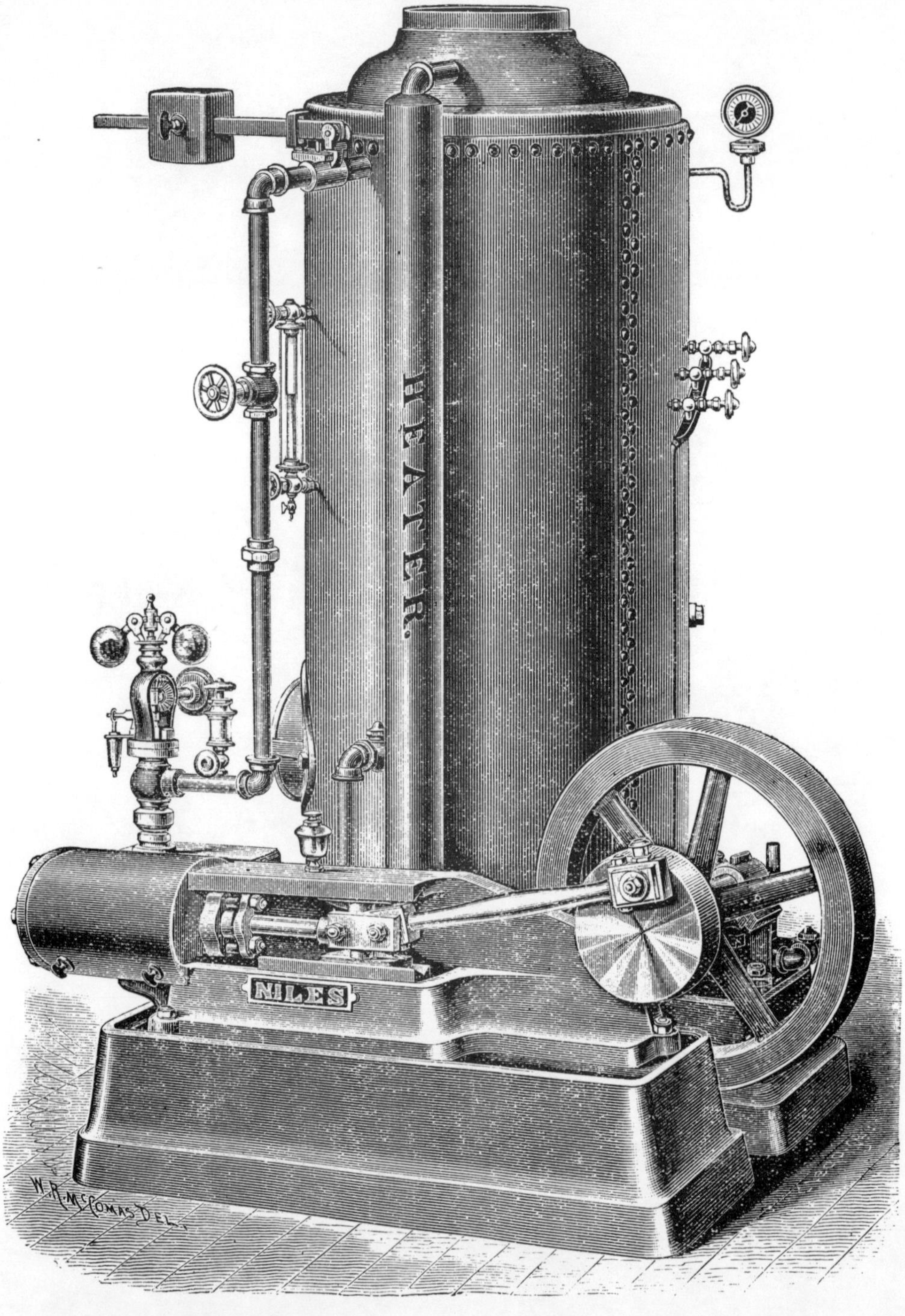

The eight-horsepower Niles engine.

Increasing use of steam for everyday activities brought forth a class of small engines and boilers, such as the Niles Steam Engine, which was also sold by Parke and Lacy. These powerplants were sized to meet any customer needs; in two-, four-, six-, eight-, ten-, and twelve-horsepower models. The larger units worked under sixty pounds of steam pressure. The engine and boiler were mounted together upon a substantial cast-iron base, forming a complete outfit. Boilers were made entirely of wrought-iron, and had fifteen square feet of heating surface per horsepower. The tubes were arranged below the water line to prevent the possibility of leakage. Each engine above four-horsepower was provided with a heater, into which the exhaust steam escaped; the feed water being heated to near boiling point before entering the boiler, which supposedly saved fuel, and prevented the injurious effects of injecting cold water into the boiler. Trimming was the best, comprising gauge cocks, glass water and steam gauges, safety and check valves, blow-off cock and feed pump. (*Mining and Scientific Press*, February 1, 1879)

The Niles single drum mining hoist.

Niles also made single and double drum mining hoists, to suit the requirements of mine owners. Made for both surface and underground work, these engines could be powered by either steam or compressed air. By 1879 Parke and Lacy had put several of these units in operation in California, one 8 x 12-inch pair going to the Guadalupe Quicksilver Mine, and another at Volcano, Amador County. These engines were compact, and as the gearing was all cut they were noiseless. The above drawing is of a 10 x 12-inch engine, which would hoist 2,500 pounds 1,000 feet with sixty pounds of air. (*Mining and Scientific Press*, February 15, 1879)

Parke and Lacy's mainstay as far as hoists were concerned, were those made by the Lidgerwood Company of Brooklyn, New York, which produced both large and small units. Shown at left is the Lidgerwood portable hoisting engine, adaptable to mines, quarries, pile driving, and moving loads on wharves. This hoist was of the smaller or dummy locomotive type, with all connections, valves, steam passages, pistons balanced and fitted for rapidity of action, economy of fuel, and durability. Steel was used throughout for piston rods, valve stems, cross heads, crank, valve and pump rod pins.

The boiler was made of the best brands of iron and tubes, and came with the most improved gauges, cocks, stop, check, blow and throttle valves. The pumps were simple and reliable, driven direct from the crank shaft by an eccentric and long connection, and provided with drainage and air cocks for relief and cold weather uses. Parke and Lacy sold a number of styles of these portable engines through their subsidiary of Berry & Place Machine Company. (*Mining and Scientific Press*, August 4, 1883)

Miles B. Dodge invented this ore pulverizer about 1880 and tested it in California mines for two years. In 1882 Parke and Lacy marketed it to the mining world. According to advertisement, the Dodge pulverizer produced no slime, retaining ore values. When set up in a mill, said this illustration's caption, "the Dodge machine worked in concert with a rock breaker. A thousand pounds of the large pieces of hard ore, selected to grind the ore as it comes from the breaker, are put in at *E*, the opposite end of the machine from the breaker, and more added as these wear away and are ground up. The machine being polyhedral in shape, there is little wear upon it. The rock does not slide in mass, as in a cylindrical machine, but falls over at each angle upon itself, thus throwing the wear upon the ore instead of the machine. Whatever wear does come upon the pulverizer, it is on the steel lining of the grate bars C and check plates *D*.

"The grate bars are held in place by the plates *D*, and wedges, *K*. As the ore is pulverized it passes through the grate bars to the screens, *F*, the fine powder sifting through into the box, *H*, or into covers, and the coarser particles coming back into the pulverizer through the openings under the grates at wedges, *K*. The machine must be run a certain way to allow these coarse particles to come back into the machine and be ground finer. The wooden housing, *G*, effectually confines the dust in the machine. There are tight and loose pulleys at *I*, which turn a pinon geared into the large spur wheel at twenty-five revolutions per minute. *J* are the bearings in which the pulverizer rests. Three sizes of pulverizers were made: three feet, four feet, and six feet in diameter. The six foot machine is so constructed that water can be used in pulverizing. The bars which form the interior of this pulverizer are shod with steel on the wearing surfaces so that they last a long time. The peculiar angulated shape of the pulverizer induces constantly recurring falls or drops of the ore, or that there is a successive breaking and grinding action, very effective in reducing the ore." (*Mining and Scientific Press*, December 16, 1882)

Notwithstanding the fact that direct-acting pumps for draining mines cost less than any other style of hydraulic equipment, a prejudice long existed against their employment on account of concussion, which was destructive to pipes and connections. It was claimed the Deane Mining Pump , introduced to Western mining in 1879 by Parke and Lacy, eliminated this by applying a simple cushion (first used on these pumps) to the steam cylinder, compelling the piston to stop and start slowly at the end of each stroke, so that the water valves had time to seat quietly, thus eliminating shook or jar. The illustration at left shows a Deane pump, piston style, with 12-inch steam cylinder, 7-inch water cylinder, and 12-inch stroke. These piston mining pumps were designed for situations where water was free of grit. They were lighter, more compact, and cost less than plunger pumps of equal capacities. Below: is a Deane plunger pump with 14-inch steam cylinder, 8-inch plunger, and 12-inch stroke.

The Deane system was advertised as being "the simplest; as the parts peculiar to it are less in number than those of the Cameron, less than one half those of the Knowles, and about one-third those of the Blake." (*Mining and Scientific Press,* January 18, 1879)

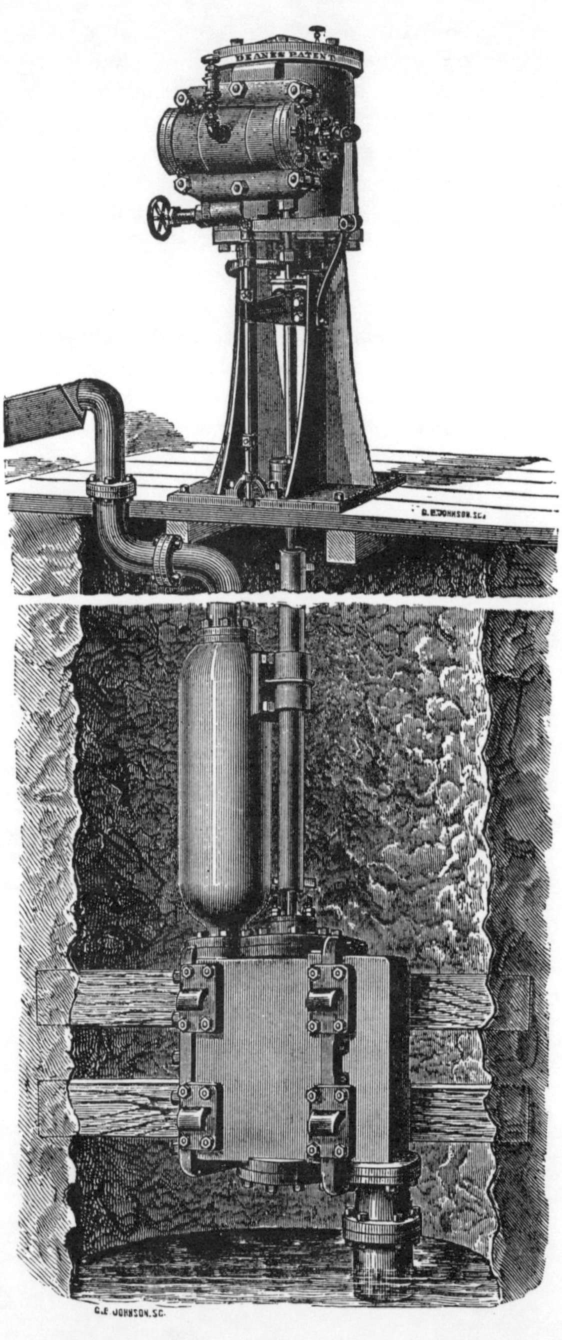

Parke and Lacy also stocked the Deane double acting vertical mining pump and engine. The engine was placed over the mouth of the shaft or well, and was connected with the pump at the bottom by a guided rod, and one pipe, that for discharge of water. These machines were made with any diameter of cylinders, from small to large, and with any desired length of stroke. The above engraving represents the double acting vertical pump and engine, 16-inch steam cylinder, 10-inch water cylinder, and 24-inch stroke. (*Mining and Scientific Press*, January 1, 1879)

Parke and Lacy installed this Cope & Maxwell pump in the Yellow Jacket Mine, on the Comstock, in 1881. The weight of the pump was about 3,400 pounds. It was lowered to the sump of the shaft by means of wire rope. The air, steam, and water pipe were made in equal lengths, so they could be simultaneously lengthened as the pump was gradually lowered. The pump had a 12 x 18 steam cylinder, and a 7 x 18 plunger, with a capacity of about 20,000 gallons per hour. It could be run either by compressed air or steam. (*Mining and Scientific Press*, June 25, 1881)

At left: The Cyclone Crusher, better known as Wall Crushing Rolls, was patented and developed by E. A. Wall to crush silver-bearing sandstone at Silver Reef, Utah. According to Wall, "it differs from the Cornish and other roller crushers in common use, chiefly in the novel construction and arrangement of the crushing faces, which consist of series of parallel corrugations, extending across the face of the shells, either parallel or inclined to their axis, the corrugations being rounded or curved with such proportions, that when intermeshed and rotated, every portion of the surface of each will press equally upon the counterparts of the opposite roller." For pan or plate amalgamation, it was claimed that three pairs of these rolls would give a greater daily production than a battery of 40-stamps. Parke and Lacy distributed these machines throughout Utah, Montana, and Australia. (*Mining and Scientific Press*, February 4, 1888)

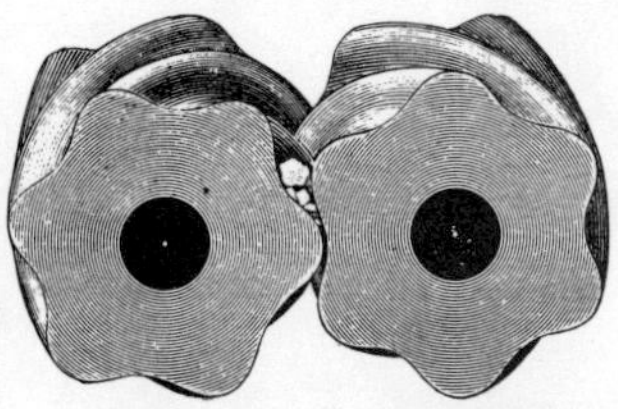

Fig. 2.—THE WALL ROLLS.

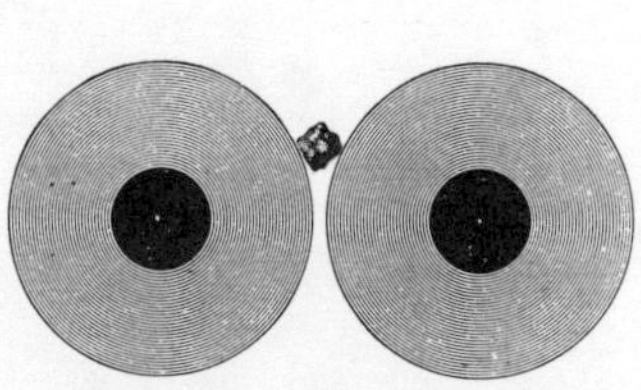

Fig. 3.—THE CORNISH ROLLS.

At 1880 many Western mills were driven by water power, and water wheels were available to suit almost any need. Parke and Lacy offered the Globe Case (at right) which was a horizontal mining wheel which accommodated an extremely small inlet and head-pipe. It was designed for mining purposes and for small wheels under high heads, where use of gears was difficult or impractical. (*Mining and Scientific Press*, June 19, 1880)

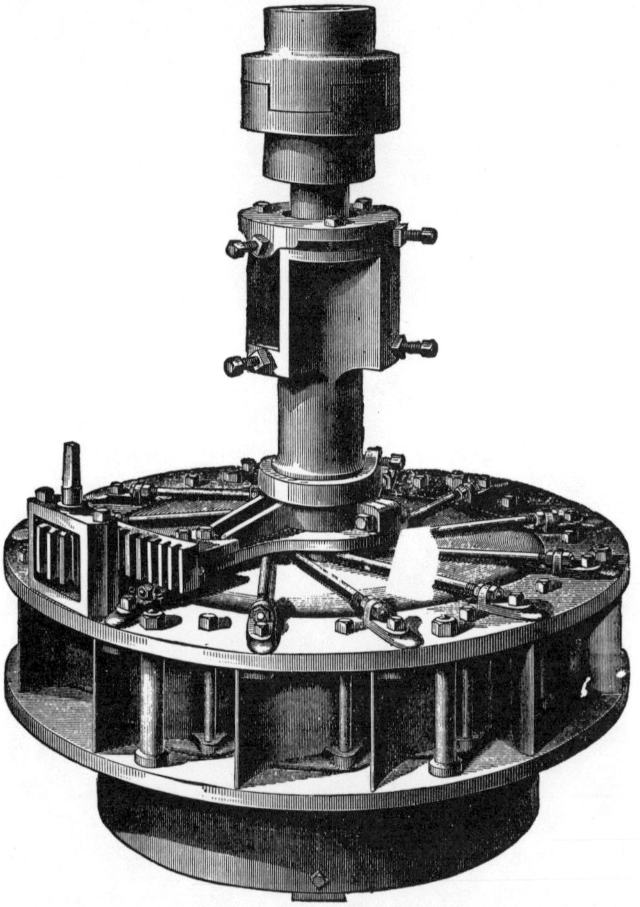

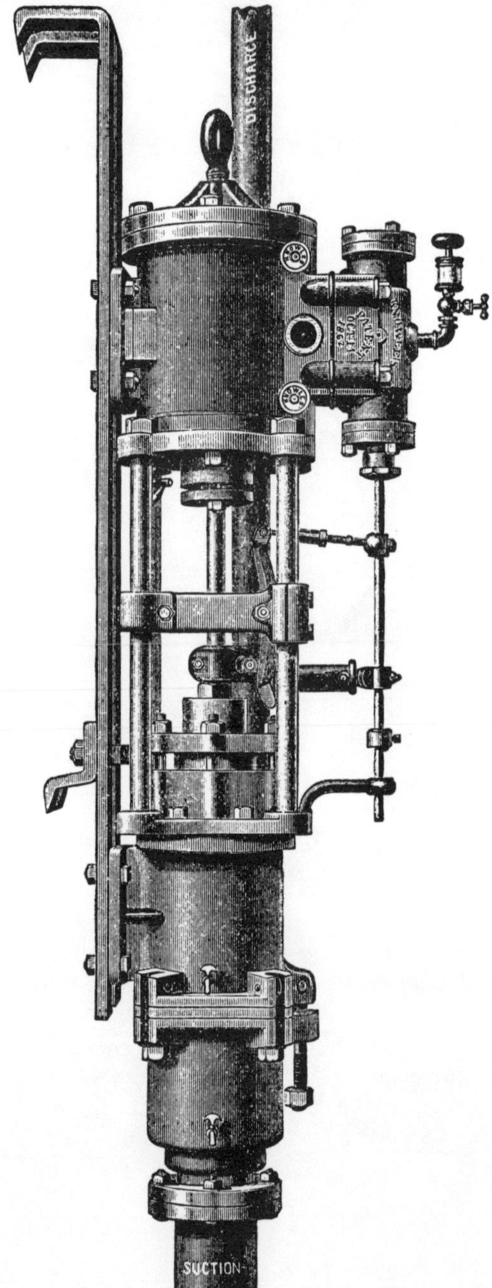

Parke and Lacy introduced just about every variety of Eastern pump to West Coast mining. In many instances they coupled the Leffel Double Turbine Water Wheel (above) with the Globe Case. In 1880 they sold various sizes of the Knowles' Vertical Sinking Pump of the bucket plunger pattern. Shown at right is the model with a 10-inch cylinder, five-inch plunger with 10-inch stroke. Six feet long, it occupied a space in the shaft $22\frac{1}{2}$ by 15 inches, and weighed 1,000 pounds. The dogs shown in the engraving hooked to shaft timbers and supported the pump. The lower dogs were adjustable. (*Mining and Scientific Press*, November 20, 1880)

Sydney, Australia, Portland, Oregon, Salt Lake City, Utah, and Butte, Montana. The latter two offices did business under the name of Utah & Montana Machinery Company. In early February 1889 Lyman Parke retired from business. The old firm was dissolved and a new company incorporated with $400,000 in capital stock subscribed as follows: L. C. Parke, $199,000; B. T. Lacy, $199,000; J. O. Harron, $1,000; A. F. Price, $500; and M. S. Norton, $500.[25]

L. C. Parke died in San Francisco on November 11, 1899. Benjamin Lacy remained as company president until his death on May 21, 1901. Shortly thereafter,

The first application of the Pelton Water Wheel was in 1881 at the Idaho Mine in Grass Valley, California. The wheel's performance was so ahead of its competitors that it was not only utilized at the mine, but soon gained popularity throughout the Mother Lode region. (*Mining and Scientific Press*, May 26, 1888)

the firm was taken over by J. O. Harron, mining engineer Thomas Rickard, and A. J. McCone, a vice president of the Fulton Engine Works of Los Angeles and the Fulton Foundry of Virginia City. The firm operated under the name of Harron, Rickard & McCone for another decade, with offices in San Francisco and Los Angeles.

The Pelton Water-Wheel Company was organized early in 1888 by engineer L. A. Pelton. Located at 127 First Street, San Francisco, the company specialized in water wheels as motive power for mining and milling enterprises. A. P. Brayton, a proprietor of the Pacific Iron Works for thirty-five years, joined Pelton in 1890 as sales manager.

L. A. Pelton was an inventive genius, who after years of persistent experimentation, successfully harnessed in 1881 the power of water with a "hurdy-gurdy" tangential or pressure water wheel. The first application of the Pelton water wheel, however, was in 1884 at the Idaho Mine in Grass Valley, where tests were run on the principal wheels being developed at the time — the Donnelly, the Knight, and the Pelton — to determine their relative value in powering mining operations. Brake tests proved Pelton's wheel far superior to the others, and sixteen of them were put in use running mills, hoists, pumps, and air compressors. During all of this period there were no breakdowns of Pelton wheels, although hoisting was carried on from a depth of 1,600 feet. The advantage of the Pelton wheel as motive power for mining was proven. The Chollar Mine at Virginia City, Nevada, soon adopted the device, installing one on the Sutro Tunnel level of their shaft. Running under a 1,500 foot head, the wheel drove five dynamos, of 100 horsepower each to generate electricity to power a surface mill five miles away.

Other mining companies soon adopted Pelton motive power, among them the Adams Mine at Rico, Colorado; two Grass Valley mines — the North Banner and the Pittsburg — each took four wheels; and other Pelton wheels went to the Gold Bank Mine in Butte County; the Hathaway Mine, Placer County; the Golden Gate Mine, Tuolumne; the Tiger Mine in Canyon Creek, Idaho; and the Central American Reduction Company and the Olancho Mines, both in Honduras, Central America.[26] Early in 1890 the Japanese government purchased an eight-foot Pelton wheel to operate a hoist at one of its coal mines.[27]

In 1889 Pelton provided the Treadwell Mill of Douglas Island, Alaska, with a plant capable of furnishing 500 horsepower, enough to drive 240 stamps, ninety-six concentrators, and twelve ore crushers. This plant consisted of a single Pelton wheel, seven feet in diameter, operating under a head of 490 feet (630 cubic feet of water per minute), making 235 revolutions per minute.[28]

By 1887–88 Pelton Wheels were being used to generate electricity not only for mining but for municipal use.[29] Under personal supervision of A. P. Brayton, six wheels were installed at the Roaring Fork Electric Power Company of Aspen, Colorado.[30]

The Knight Water Wheel, patented in January 1875, and used throughout Amador County, California, offered Pelton stiff competition. By the mid-1880 Pelton had conquered his opposition and was linking his wheel to the equipment of major manufacturers. Below: A pair of Ingersoll-Sergeant compressors driven by Pelton Water Wheels. (*Mining and Scientific Press*, June 30, 1894)

Pelton designed some unique machinery during 1894. Its wheels were harnessed to a pair of Ingersoll-Sergeant compressors destined for a mine in the heart of the Andes, a hundred miles east of Lima, Peru. The entire plant was made in sections for mule packing, no piece exceeding 300 pounds in weight.[31] At the same time, the company built for the Milwaukee Mine at Coeur d'Alene, Idaho, a water-powered hoist embodying all safety features known at the time. Powered by six-foot diameter wheels, running under 850 feet head at 375 revolutions, the hoist could carry a load of 5,500 pounds at a speed of 400 feet per minute, and work to a depth of 1,500 feet.[32]

December 1894 was a busy month for Pelton. A dozen water wheels were sent out, including two to Petropolis, Brazil, each developing 300 horsepower under 260-foot head. With the latter order went 3,000 feet of thirty-inch pipe.[33] By 1895, Pelton wheels had been installed in almost every country,[34] forcing the company to seek larger quarters in 1899. A four-story brick building was erected on Main Street, between Mission and Howard, the shop portion of which embraced 20,000 square feet.[35] Unfortunately, the plant was destroyed by the San Francisco earthquake and fire of April 1906. Amazingly, Pelton rebuilt. Within nine months it had erected, in the block bounded by Harrison, Alabama, and 19th Street, the "first modern machine-shop to be erected in San Francisco since the fire."[36] The company persisted into the 1920s.

The National Rocker Quartz Mill was invented by Stephen Kendall who put together a one-stamp mill in 1878. Manufactured by the National Iron Works, the mill was simple in design, no more than a semi-circular mortar, fitted with removable dies, resting on rack bars. Inside the mortar was a large roller. When the mortar rocked back and forth, the roller moved over the dies, crushing the quartz. Capable of crushing ten to fifteen tons of quartz per day, the mill cost $500 with frame. Weight of the unit with countershaft, boxes, pulley, crank and connecting rod was 5,300 pounds. (*Mining and Scientific Press*, November 13, 1886)

The National Concentrator, also invented by Stephen Kendall, could be run by hand, or direct from a main shaft without use of pulley and countershaft. The machine was like an ordinary percussion table. The lower portion of the table had more of an incline than the upper, where water was applied. A semi-circular receptacle extended transversely under the upper edge of the table, over which was a perforated cover or board. Water admitted through the pipe shown in the illustration, flowed under the perforated cover, up through the perforations and evenly over the table, moving heavier particles upward, and sending lighter portions down and away. The jarring of the machine moved the heavier particles through the perforations into a receptacle. Sulphurets (pyrites) were then removed from the tube shown in the illustration. The machine cost $250. (*Mining and Scientific Press*, July 9, 1881)

The National Iron Works began operation in early November 1879. The proprietors were L. C. Marshutz, a former partner of Hinckley & Company, and T. G. Cantrell, formerly of the firm of Hawkins and Cantrell. Each man had twenty-five years experience in the iron business in San Francisco.

They leased the former premises of the San Francisco Planing Mill on the northwest corner of Main and Howard streets. The building was a large one: three stories high, covering an area 91 x 137 feet. The machine shop, located in a lower corner room, had doors opening on both streets, which facilitated handling of large pieces of work. At one corner were offices. Back of the machine shop was the foundry and engine room. The cupola furnace was of twelve tons capacity, lined with fire brick all the way to the top. The blast was furnished by a blower operated by an engine independent of other machinery, which gave the cupola man perfect control of the blast. The core furnaces were twenty feet long.

A drafting department and pattern shop, the latter equipped with band and circular saws, and lathes, were on the second floor. The foundry had a large collection of patterns and drawings, purchased from the defunct Hope Iron Works. The third story was used for a pattern room. The plant's equipment, consisting of several lathes, planers, drill presses and slotting machines,

At left is a portable hoisting engine built by the National Iron Works in 1880. The company made a specialty of such hoists, producing them in several sizes. The largest size weighed 8,000 pounds, and had an engine with a 8 x 12-inch cylinder, capable of lifting 1,000 feet. The smallest size, 6 x 12, would hoist 500 feet, and weighed 5,000 pounds. The same style engine was also made with double cylinders. They were advertised as being "made strong, plain and neat."

These hoists had one advantage: they were shipped from the foundry ready to be put to work as soon as they arrived on the ground. No skilled labor was required to get them going. "A cheap and good rig for opening a mine," these portable hoists were shipped throughout California, Nevada, Lower California, and Sonora. (*Mining and Scientific Press*, December 18, 1880)

was considered first class, all having been made by Bement & Sons of Philadelphia.

The shop's stated specialty was the manufacture of engines, flour and sawmill irons, all "at a low figure," as well as quartz crushing machinery.[37] Of the latter were two devices invented by Stephen Kendall, an Amador County millman. In 1878 he patented the Kendall Quartz Mill, a single stamp mill, fabricated at the Aetna Foundry. Upon collapse of Aetna, National produced the mill in a number of models. They were built ready to use, no millwright being necessary to set them up. A few years later, Kendall introduced the "National Rocking Battery," a simple, portable quartz mill that was no more than a semi-circular mortar resting on rack bars placed on timbers, so that it always maintained a level position while in motion. The mortar, fitted with removal dies, was a large roller with a circular slot in its center into which fitted a central rib which prevented the roller from sliding side-to-side. The mill was rated at ten to fifteen tons per day, and its price, without frame, was $500.[38]

National also fabricated portable hoisting engines in several sizes. Like their quartz mills, these hoists were advertised as self-contained units, "ready to be put to work as soon as they arrive on the ground." There must have been truth in National's advertising, for these hoists were widely used throughout California, Nevada, Arizona, Sonora, and Lower California.

On September 8, 1888, the firm was destroyed by fire, throwing 200 men out of work. The disaster swept away years of toil and industry. With indomitable pluck and energy of old Californians, the owners rebuilt in five months, and from a massive, fireproof brick building, resumed construction of engines, hoists, and mills.[39] The company, however, did not weather the decline of San Francisco's equipment trade in the 1890s and vanished from the foundry scene by 1895.

The Vulcan Foundry and Iron Works at 137 and 139 First Street, corner of Natoma, was established about 1861 by George Gordon and E. T. Steen, both affiliated with the California Sugar Refinery. They did not stay with the foundry, for in 1864 the owners were listed as Charles Steiger, Samuel Aitkins, Richard Ivers, and Paul Torquet, the latter serving as manager. Torquet died, and his interest was assumed by N. D. Arnot. Well managed, the company grew rapidly. Within three years it employed 160 men and consumed an average of 176,000 pounds of iron a month for castings, 44,000 pounds of boiler iron, and 37,000 pounds of wrought iron. Like all San Francisco foundries, Vulcan turned out quartz milling equipment: stamps, separators, concentrators, and amalgamators. Of the latter, was the Hepburn and Peterson amalgamator, widely used in Washoe.[40]

Vulcan excelled at designing steam engines, and it turned out monstrous ones. In 1863 it built an engine for the Gould & Curry Company's mill at Virginia. Working at its intended speed of twenty-five revolutions per minute, under 100 pounds per square inch of steam, the engine was rated at 260 to 300 horsepower, capable of running 150 stamps with full compliment of pans. It was the largest stationary engine built in San Francsico up to that time.

An idea of the size and proportion of this engine can be formed from the fact that its main shaft weighed 7,000 pounds, the bed-plate 14,000 pounds, its beam 7,500, its cylinder 4,200, and the entire engine exclusive of boilers, weighed eighty tons. Its cost was $30,000, excluding boilers and shafting. The *Mining and Scientific Press* said that the engine, while substantial, had an "elegant and artistic" appearance.

"A massive bed-plate, 21 feet long and 6 feet wide forms the base of the structure. The cylinder is securely bolted to one end of this bed-plate, and is itself a fine piece of work, 30 inches in diameter and 6 feet stoke. At the opposite end is the main pillow block, while from the center and from each end of the bed-plate, elegantly moulded columns, supported by diagonal braces, all in gothic style, rise to a height of 17 feet, to support the beam centers and sustain a neat entablature extending around the upper portion of the engine. Brackets are bolted to this entablature, upon which a light platform will be placed, surrounded with iron standards and a rail or guard of polished brass. . . . The engine beam is of an approved pattern, 15 feet long and 7 deep. The connecting rod is 15½ feet long, of wrought iron, with the usual braces. It is a handsome piece of forging and neatly finished. The crank is also of wrought iron, as is the shaft; the latter being 12 inches in diameter and 14 feet long."

The engine was completed in February 1864, and formed "quite a card" for the Vulcan Iron Works. The engine was built under the supervision of R. G. Carlyle, chief engineer for the Gould & Curry Mining Company. Paul Torquet, manager of Vulcan, and proprietors C. R. Steiger, R. Ivers, and S. Aitkens took a personal interest in the mammoth engine, as did Joseph Moore, the machine shop foreman. The engine was designed and drawn by Frank Euler, Vulcan's draftsman.[41]

Vulcan also built another large engine for the Yellow Jacket Mine, with a 22-inch diameter cylinder, and a six-foot stroke, capable of driving an 80-stamp mill. Other smaller engines were shipped to the Mina Rica Mill, West Point, Calaveras County, California. One of the most interesting orders turned out by Vulcan was a complete quartz mill for a Japanese mining company. It consisted of a 10-stamp mill with amalgamating appliances. The motive power was a large overshot water wheel which the Japanese built at the mine. The battery was a substantial one, with iron columns and head pieces, iron being used wherever possible. The two batteries of five stamps each, weighing about ten tons, were set up and proved before shipment. Weight of the entire consignment was about thirty-five tons.[42]

In 1870 Vulcan did a large amount of work not related to mining: large retorts for the Pacific Wood Pre-

Fig. 1.

Fig. 2.

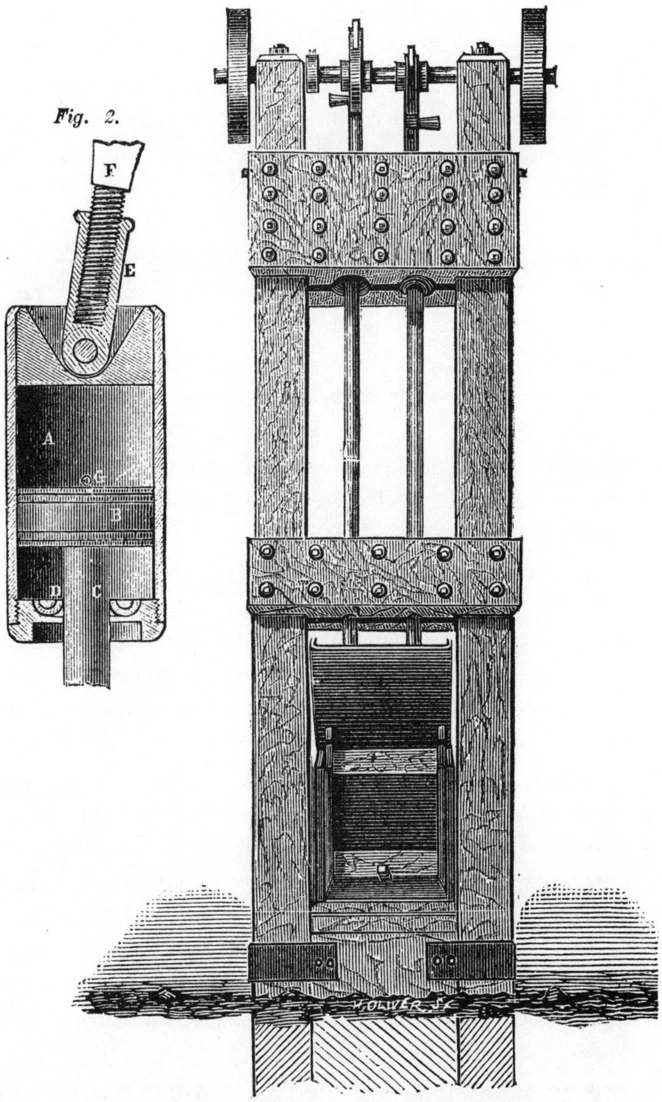

The Reliance Ore Battery (Figure 1), devised by E. T. Steen and manufactured by the Vulcan Iron Works, was constructed for high speed running. This battery had no cams and tappets, each stamp running from 150 to 400 strokes per minute. Stamp stems were connected to an air cushion (Figure 2), which dampened the impact of the stamps and prevented the machine from shaking itself apart. Elimination of rigid connections allowed high speed running, and it was said the mill would accommodate itself to any amount of ore in the battery. The battery was quite portable; the weight of all iron and wood work, except the mortar block, for a battery of two stamps of eight inches diameter, being only 5,400 pounds. Three sizes were made, which sold for $600 to $1,200, including a Victor ore feeder; the whole being ready to set up.

Taking rock up to two and a half inches, this mill could be run by any kind of power, and used either as a prospecting mill, or a full working crusher. (*Mining and Scientific Press*, October 2, 1880)

serving Company and the California Sugar Company, as well as fabrication of machinery for the San Quentin state prison.[43] In 1872 Vulcan enlarged its plant and embarked upon a program of building hoisting equipment for deep mining.[44] Like other foundries, Vulcan turned out massive machinery for Washoe mines. When the Comstock faltered in the late 1870s the foundry switched to construction of lead furnaces. In fall of 1893 it furnished the Germania Lead Works of Salt Lake City, Utah, with two Bruckner furnaces. These were large units, the cylinders being 102 inches in diameter by twenty-two feet long, capable of receiving twelve-ton charges. Each cylinder weighed twenty-three tons.[45] At the same time, it built a leaching mill for the Ontario Mine at Park City, Utah, as well as furnishing stamp mills for the Jones Mine in Shasta County, California.[46]

In late October 1880 Vulcan unveiled the Reliance Ore Battery, a mill devised by E. T. Steen, a former partner in the company, which was intended to revolutionize ore dressing. The mill did away with cams and tappets, the stamps being connected directly to the drive shaft. Such an arrangement, so said the advertising, allowed for high speed running, "any desired number of blows per minute being possible, a result which cannot be obtained with the ordinary stamp battery when cams and tappets are used." The Reliance Mill was made in three sizes, with prices ranging between $600 and $1,200. It is not known how successful this battery was. Judging from lack of advertising it was probably a failure. As mentioned before, millmen were a subborn lot, and by 1880 the gravity stamp mill was the proven way of crushing rock. Which might explain another Vulcan departure.

In the early 1880s Vulcan began manufacturing ammonia refrigeration plants. At first these were primitive appliances, but scientific and mechanical progress soon rendered ice-making machines practical. In 1885 Vulcan put its first ice-making machine into the Hotel del Monte at Monterey, California. From then on, West Coast industry cried for refrigeration, until the line consumed Vulcan's full energy. Its plants went to packinghouses, cold storage facilities, ice factories, markets, hotels, public institutions, and dairies. In 1893 the company furnished the Gambrinus Brewery Company of Portland, Oregon, with a thirty-five-ton refrigeration plant, powered by a Corliss engine. Another large plant went to a salmon processing firm at Port Essington, on the Skeena River, British Columbia.[47] The company also installed refrigeration units in steamships transporting fruit, fish, and other perishable products.[48] Eventually refrigeration displaced mining in Vulcan's manufacturing agenda, and by 1900 the firm no longer produced mining equipment.

A mechanic by trade, Englishman Joshua Hendy arrived in the United States in 1836, taking up residence with his family first in Louisiana and then Texas. In 1849, at age thirty-two, he journeyed to California, arriving at San Francisco on September 19 aboard the steamship *Oregon.* Finding work in one of the city's open-air boiler shops, Hendy waited for an opportunity to put his skills to work for his own benefit. That opportunity came several years later when he learned by chance that three portable sawmills were for sale. Buying them, he placed two in operation in Placer County, and the third he took to Sonoma County, where he joined forces with the Duncan brothers, starting what eventually became the Duncan Mills. He returned to San Francisco to establish the Joshua Hendy Machine Works, at 49 and 51 Fremont Street, a sawmill manufactory.

While fabricating sawmills, Hendy took on custom foundry work, repairing quartz mill components. The multitude of bent stems, worn and chipped shoes and dies passing through his shop told Hendy that there was something wrong in the way these mills were being run, or at least, in the way they were being fed.

Hendy knew the theory of ore feeding; in fact, anyone who had an elementary knowledge of chemistry knew it. "Take an ordinary hand mortar and pestle, and attempt to pulverize a quantity of quartz; drop in but a small quantity, just sufficient to cover, in a thin layer, the bottom of the mortar, and it will rapidly pulverize under the blows of the pestle; drop in a large quantity, and the blows of the pestle will be ineffective, and time and labor lost in the effort to reduce it to powder."[49]

Millmen being a stubborn and sometimes stupid lot, failed to recognize this principal. The only "really intelligent" way of furnishing ore to the stamps, they told Hendy, was by hand feeding. That may have been true as long as batteries were small, for an alert feeder could size the ore, breaking-down big pieces with a sledge, and at the same time keep a constant depth of pulp on the dies. But as batteries were added, and mills increased production by addition of untrained labor, the factor of human error was multiplied, rendering hand feeding "irrational and uneconomical." It was hard for inexperienced, lazy, or drunken men to keep ore coming in proper proportion for the stamps to work efficiently. Too much ore and the head was literally cushioned, reducing the output, which often

"THE $1,000 CHALLENGE"

Ore Feeder for Quartz Mills.

OVER 800 ARE NOW IN USE, GIVING ENTIRE SATISFACTION.

Awarded First Premium at the Tenth and Twelfth Industrial Fairs of the Mechanics' Institute.

Twenty Per Cent. More Ore Crushed with Fifteen Per Cent. Less Wear of Iron than by Hand Feeding.

The accompanying cut illustrates the recently introduced Grip, and also the Spring Attachment, which replaces the Weight heretofore used, and which are obvious improvements.

It is now fully demonstrated, after careful and long continued experimentation and practical use, that the plan upon which a perfect Ore Feeder must be constructed is that of a carrier, and not that of a shaking-table. Uniform and accurate feeding is not possib'e upon the latter plan. The ore must be evenly carried, upon a steadily advancing plane or table, to the line of discharge, and there simply dropped. Jerky or spasmodic contrivances will not answer the purpose for wet or sticky ores.

The Challenge Ore Feeders are now in Use in the following Mills, besides many others

Soulsby	20	Stamp	Tuolumne county, Cal.
Sheep-Ranch	20	"	Calaveras " "
Mahoney	40	"	Amador " "
Zeile	40	"	" " "
Placerville	40	"	El Dorado " "
Gross	80	"	" " "
Julian	20	"	Placer " "
St. Patrick	15	"	" " "
Providence	20	"	Nevada " "
Omaha	10	"	" " "
Green Mountain	60	"	Plumas " "
Plumas-Eureka	60	"	" " "
Bulwer-Standard	30	"	Bodie Dis., Mono, "
Standard	20	"	" " "
Noonday	30	"	" " "
Bodie	10	"	" " "
Christy	5	"	Uintah Co., Utah.
Ontario	40	"	Parley's Park, "
Contention	20	"	Tombstone Dis, Arizona
Grand Central	20	"	" "
Harshaw	20	"	Patagonia, "
Sunshine	20	"	Idaho Springs, Col.
Homestake	2[illegible]0	"	Black Hills, Dakota.
Father De Smet	8[illegible]	"	" "
Hidden Treasure	40	"	" "

Superiority of the "Challenge" Ore Feeder Demonstrated!

At the "Christy" Mill, Uintah County, Utah, the "Eclipse" Feeders, (conceived by E. Coleman) were introduced, but not carrying a regular supply of ore for the crushing capacity of the stamps, were replaced by the "Challenge," which are now running and the stamps crushing forty (40) per cent. more ore than was done by the "Eclipse."

The "Harshaw" or "Hermosa" Mill, of Patagonia District, Arizona, was also originally fitted with "Eclipse" Feeders, but after a few weeks trial they were pronounced inadequate to the work, discarded, and the "Challenge" adopted.

The "Silver King," Mill of Arizona, also removed the "Eclipse" Feeders to give place to the "Challenge."

The "Sola" Mill, of Brown's Valley, Yuba County, Cal., was fitted with "Victor" Feeders, manufactured by E. T. Steen, but proving insufficient, the "Challenge" Feeders were substituted.

Four of the "Victor" Feeders, manufactured by E. T. Steen, were also placed in the "Alexander" Mill, at Grantsville, Nevada, but after a fair trial were discarded, and Hendy's Feeders fitted, and four others of the same pattern added when the second twenty stamps were erected.

These cases are simply cited from among many similar instances, in proof of the vast superiority of the "Challenge" Feeders over all others.

Mr. H. W. GREENWELL, of the Tingman Mine, El Dorado Co., or any other person can make $1,000 by disproving the acknowledged superiority of my "CHALLENGE" Feeder, in a competitive trial, side by side, with any other Feeder, on wet or sticky ores from any mine, and I will give $100 to any one who will procure the acceptance of this challenge, and arrange for a fair and impartial test on merit.

I mean business, Gentlemen, "either put up or shut up."

JOSHUA HENDY,

Manufacturer of all descriptions of Quartz and Saw Mill Machinery. A Large Stock of both New and Second Hand at Machine Works. Nos. 49 and 51 Fremont Street, San Francisco.

Agent for "BAKERS" Rotary Pressure Blowers.

Joshua Hendy's ad from the March 19, 1881, *Mining and Scientific Press*, offering $1,000 to anyone who could disprove the "acknowledged superiority" of his Challenge Ore Feeder. By 1881 there were over 800 Hendy feeders in operation throughout Western America. Nevertheless, Hendy faced competition in the ore feeder market, which he eliminated by buying the patents and distribution rights to his competitors' machines.

Joshua Hendy patented this ore feeder on June 24, 1873. The important feature of the mechanism was that it was a carrier feeder, the ore being carried over a roller to the line of discharge, which was an essential element in construction of any automatic ore feeder. The rate of ore flowing over the roller was controlled by an adjustable gate.

resulted in the head becoming detached from the stem. On the other hand, if feeding of ore was too slack, the dies were exposed, the concussion intensified, causing bent or broken stems, and battered shoes and dies. In either case, the result was the same: costly wear and imperfect reduction of ore, or both.

Hendy believed a device could be designed to keep the depth of pulp on the dies constant throughout the battery, as long as work was carried on.[50] What he came up with was the "Challenge" Automatic Ore-feeder, patented on March 17, 1874. The device was no more than a pyramidal hopper with an inclined floor, connected with the feed-opening of the mortar-box by a spout or chute. A bar attached to the tappet of the mill actuated the feeder. Adjustment of the rod one way and the stroke of the tappet caused the device to "feed low"; adjust the other way and the feeder allowed in more ore for a "high feed."

The Challenge Feeder was truly a labor-saving device. But selling it to the milling profession was uphill work. Millmen were resistent to change; back-breaking shoveling of ore into batteries was the tried and true method. Hendy was as stubborn as any millman, however. He persistently pointed out how his ma-

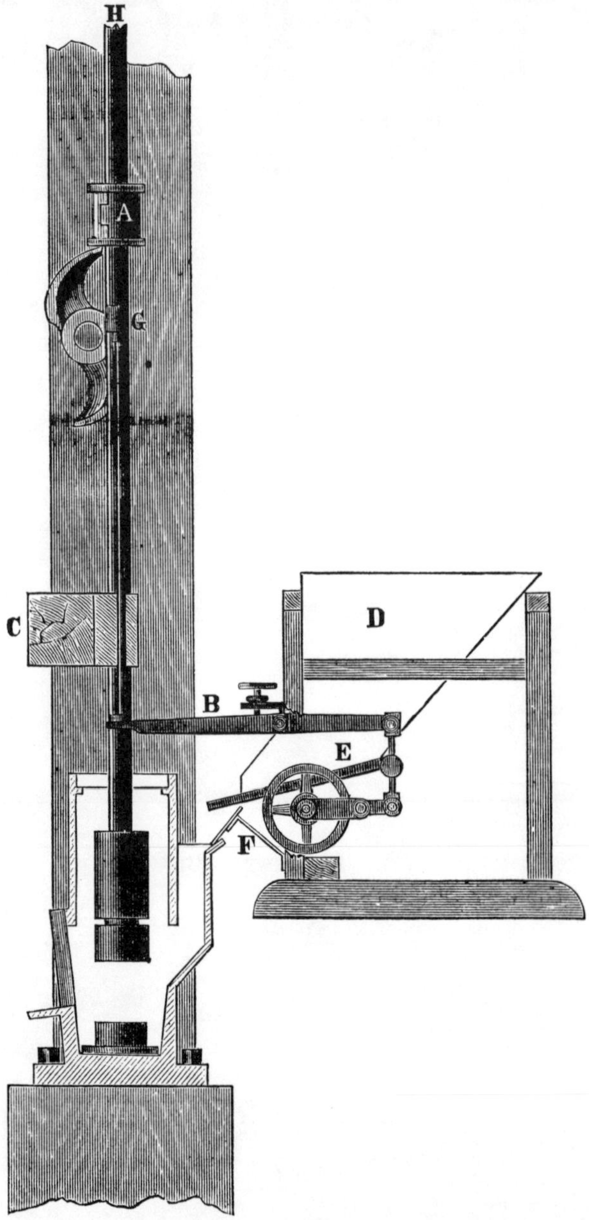

Before introduction of automatic ore feeders, batteries were fed by hand, which required constant attention to prevent the stamps from running out of ore. If that happened, the stamps would pound upon the dies and damage would occur. The above device prevented that. The ore feeder was no more than a hopper, *D*, placed in front of the mortar, into which ore was poured after it had been reduced to a desired size by a rock breaker. Being an inclined chute, the hopper fed the ore to the batteries by means of a rotating table, *E*, actuated by a lever, *B*, attached to the stem of the stamps. Thus the machine was automatic in its action; that is, the requirement of the stamps was determined by the amount of drop or force of the blow delivered by the tappet, and the amount of drop regulated by the requirements of the stamps. (*Mining and Scientific Press*, April 22, 1876)

chine could save labor and money. The merits of his feeder were undeniable. Twenty percent more ore could be crushed with at least fifteen percent less wear of iron than by hand feeding.[51] In the end Hendy conquered his opposition.

Joshua Hendy perfected the Challenge Feeder over a period of ten years, obtaining patents for improvements on December 28, 1880, and July 21, 1885. By 1881 Hendy had installed 800 Challenge Feeders in the "best appointed" quartz mills in the United States and elsewhere.[52] In September 1882 Hendy incorporated his business with himself as manager, and his nephew, Samuel J. Hendy, as president. By then other ore feeders had appeared on the market. In 1870 C. P. Stanford of San Francisco brought out a feeder, and three years later Thomas A. Cochrane introduced a belt carrier feeder. The Tulloch Roller Feeder followed about the same time. All three were widely accepted in quartz mill in Tuolumne, Placer, and Amador counties.

These feeders, particularly the Tulloch and Stanford machines, threatened to undermine Hendy's business, something he did not tolerate. Shrewdly, he met the threat head-on, buying his competitors' patents. By the mid-1880s he controlled the lucrative ore-feeder business, manufacturing exclusively the Tulloch and Stanford feeders. But it was his Improved Challenge feeders that Hendy pushed the hardest.

James Tulloch patented the Victor Ore Feeder (above) in August 1876 through the *Mining and Scientific Press* Patent Agency. The feeder was introduced to the public by E. T. Steen. Like every other feeder, the dropping of the stamp kept ore flowing to the battery. This feeder was fitted with rollers so that it could be moved easily. It measured five feet from the floor to the top of the hopper, but could be made higher or lower if desired. The ore was kept loose in the hopper. The mechanism was operated by a wedge on the lower end of the rod, shown in the engraving. As the stamp dropped, the wedge forced back a horizontal rod, which pushed against the lower front edge of the lower hopper, as it came forward again by its own weight, pushing the ore before it into the battery. The feed was regulated by inclination of the lower hopper. (*Mining and Scientific Press*, December 29, 1877)

At right is another ore feeder that appeared two years after the Tulloch, patented by James D. Cusenbary and James A. Mars of San Francisco. The feeding cylinder was secured on a foundation of timbers, which in turn were mounted on rollers so the cylinder and frame could be moved about as desired. The cylinder was made of cast iron, its outer surfaces formed into chambers or depressions separated from each other by longitudinal partitions. The cylinder and its carriage, when in working position, were placed below the hopper, so that ore from the hopper fall into the chambers, and as the cylinder revolved, it was carried along an inclined apron to the stamp below. (*Mining and Scientific Press*, August 16, 1878)

Left: The Stanford Ore Feeder. Acquiring rights to this machine, Joshua Hendy improved the device until it was as good and durable as his Challenge Ore Feeder. Selling for $75, it was the cheapest ore feeder on the market. The essential feature of the Stanford Feeder was its carrier or feed table, supported by side journals on the frame, which was actuated by rod and lever engaged to the stamp stem tappet. The best ore feeders were those that kept the smallest amount of ore under the stamp without iron-upon-iron contact. The stamps would then do the largest amount of work. A good ore feeder allowed twenty percent more ore crushed over hand feeding, and fifteen percent less wear of battery iron. (*Mining and Scientific Press*, January 31, 1880)

The feeder at right was patented in 1874 by James Tulloch, of Sonora, Tuolumne County, California. Tested at the Buchanan Mill, the Shawmut Mine, and the Knox & Boyle Mill, all in Tuolumne County , the feeder gave good performance. The Tulloch Feeder operated like all the rest. A vibrating tray placed beneath a hopper, actuated by a lever and arm attached to the tappet stem, allowed passage of ore into the battery. (*Mining and Scientific Press*, July 4, 1874)

Above: Hendy's improved Prater Concentrator, introduced in the mid-1860s, was built at the Golden State and Pacific foundries. It worked on a centrifugal principle, crushed ore being fed through the hopper, *C*, to a distributor, *B*, which deposited the sand about the periphery of the pan. Water was discharged at the center of the machine, while the sulphurets (pyrites), gold, and amalgam were taken out through a discharge gate, *E*, and caught in the box *G*. (*Mining and Scientific Press*, February 3, 1866)

Above: The Hendy Ore Crusher, introduced by Joshua Hendy in 1880, was a simple machine. Only the outer jaw moved, the inner one being stationary. There was an eccentric on the shaft on each side of the machine, and from those, wrought-iron adjustable stays reached out and supported the outer-swinging jaws, uniting on a cross bar. As seen in the bottom section, it was easy to change the machine's shoes and dies. The side plates were removed, the lower outside nuts slacked, which freed the swinging jaw. The jaw then swung over the top, leaving the stationary jaw accessible. The crusher was light, yet strong and durable; the cheapest crusher on the market. (*Mining and Scientific Press*, November 27, 1880)

Hendy also pioneered development of ore concentrators, and thousands of his Prater concentrators, manufactured at the Union Iron Works, were used until supplanted by belt machines. Undaunted by the new type concentrators, Hendy patented on July 18, 1882, the Triumph Concentrator, a belt machine which was well received, large numbers being shipped in 1887 to Idaho, Montana, the Dakotas, New Mexico, and California.[53] Hendy did not stop there.

He was also sole manufacturer, under letter of patent, of the hydraulic giant known as the Fisher double-jointed nozzle. And he was the first to manufacture the hydraulic gravel elevator, a machine which mining engineers doubted the practicality of. Hendy did not. To prove the merit of his elevator, he bought the Yreka Creek gold mining property in Siskiyou County, built an elevator and put it to work. Because the ground did not pay, this venture cost him $70,000. The machine, however, was a success, and many were put to work raising earth, sand, and gravel from shallow deposits, and working low ground where there was no "fall."

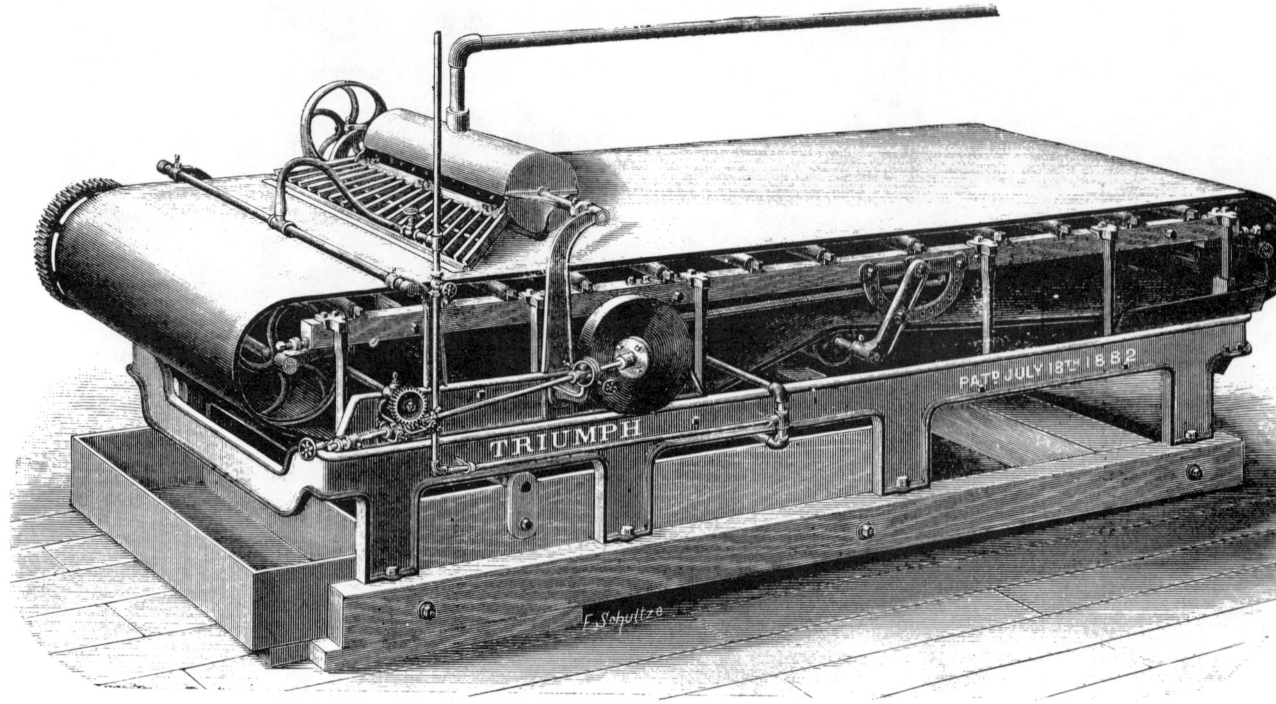

The "Triumph," an endless belt concentrator, was Joshua Hendy's answer to the Frue Vanner. Of all-iron construction, it was solid and durable, and had a steadiness of movement in production which allowed it to out perform, in many instances, the Frue machine. In tests conducted at the Original Empire Mining and Milling Company in Grass Valley, in summer of 1884, Triumph concentrators produced 21,159 pounds of concentrates, which yielded after deducting the cost of chlorination, $1,713.22 net, while Frue vanners yielded 18,180 pounds of concentrates, with a net value of $1,514.07. (*Mining and Scientific Press*, August 16, 1884)

The Joshua Hendy Iron Works also built complete gold concentrating mills, a good example of which was the 1888 construction of a 20-stamp mill at the Boaz Gold Mine, sixty miles from Phoenix, Arizona.[54] As agent, the Hendy Company dealt in engines, boilers, pumps, blowers, and machine tools.[55] The real specialty of the Hendy Iron Works were lightweight, inexpensive prospecting equipment, such as the Davis Horse-Powered Hoisting Whim. This all-steel whim was advertised as being superior to other forms of whims whose construction led to failure.[56] Hendy also manufactured a gold pan having a flange or "protector" around one side of the pan "making it impossible to lose the gold."[57]

On October 19, 1891, at the height of his career, Joshua Hendy died. He was seventy-four years old. His nephew, Samuel, took over the company and it prospered. While other foundries struggled to stay alive, the Hendy Works kept busy, moving quartz mills, hydraulic, and inexpensive equipment.[58] The company did suffer a set back in the San Francisco earthquake of April 18, 1906, when the subsequent fire wiped out its shops. The company rebuilt its offices and showrooms on the old site at 75 Fremont Street; the first permanent building erected in that part of the burned district. At the same time, the company acquired land at Murphy's Station, now known as Sunnyvale, in Santa Clara County. There, it erected machine and blacksmith shops, and carried on into the 1920s supplying equipment to the mining world.

Below: The Hendy Improved Gold Pan had a flange running half way round the pan, about an inch below the lip, which according to advertising, made it impossible to lose the gold. (*Mining and Scientific Press*, June 4, 1888)

THE HOISTING DRUM AND GEARING.

In the initial stages of developing a mine, one of the most important factors in calculating expense was the cost of power required to sink to a given depth, or until such depth is attained as to justify procurement of a steam or water-powered hoisting plant. Joshua Hendy's answer was the Davis Horse-Power Hoisting Whim, shown above. This whim was built entirely of iron and steel, mounted on a heavy base plate, so that it was not affected by cold or heat. Both hoisting and lowering was regulated by one lever; the operation so simple that it could be handled by any inexperienced person.

These whims were light and compact, weighing 1,200 pounds, and could be packed for transportation by mule. They were made in two models: the No. 1, with a capacity of 800 pounds, capable of hoisting seventy-five feet per minute; and the No. 2 of 500 pounds, hoisting 125 feet per minute. (*Mining and Scientific Press*, May 4, 1889)

Above: The Hallidie Patent Wire Rope-Way for transporting ore from mine to mill. One can readily see the advantages this method possessed in mountainous regions, from the fact that the system required neither expensive machinery nor roads; that it could be run at all seasons of the year, even when there was great depth of snow on the ground; that it could be cheaply constructed in the worst possible country; and that where there was sufficient grade not only did it run itself by gravitation, but produced a motive power at both ends of the line. Running at 200 feet per minute, the cars were carried down on one side of the posts and up the other, 240 per hour, each delivering 100 pounds of ore or twelve tons per hour, and they could run day and night continuously. Total cost of laying down the endless way was from $6,000 to $12,000 per mile, and the cost of carrying the ore was about twelve cents per ton per mile. (*Mining and Scientific Press*, September 5, 1874)

CHAPTER FIVE

Andrew Smith Hallidie and the California Wire Rope Company

WIRE rope contributed more to the advancement of mining than any 19th Century invention. Harnessed to steam engines it allowed hoisting of great loads from depths never permitted by hemp rope. Its high tensile strength assured an unmatched degree of safety, which accounts for manufacturers of mining equipment everywhere becoming dependent upon this component for all deep mining hoists, tramways, and numerous other applications. Before looking at the supplier of wire rope for Western American mining, we need a little background as to its origin.

Wire rope is generally considered a recent invention, a product of modern skill. The English claim its discovery, as do some Americans. Actually its manufacture is a rediscovered art of some antiquity, for the Italians were weaving metal wires at a time when the English were painting themselves blue. Two thousand year old bronze wire rope was dug from the ruins of Pompeii, and it is conceivable that Egyptians, who produced metal wire, were weaving it into rope at a much earlier date.[1] Wire rope-ways were used to transport goods during the Middle Ages; they are portrayed in Renaissance books; and Wybe Adam, a Dutch engineer, constructed an aerial wire rope-way at Danzig in 1644.[2] The first practical application of wire rope, however, was in the construction of a suspension bridge at Geneva, Switzerland, in 1822, and a similar bridge at Freiburg, Germany, in 1835.[3]

The first commercial application of wire rope as far as the English are concerned, grew out of the Russo-Turkish War of the late 1820s, which cut off Britain's access to hemp and forced the Royal Navy to find a substitute for naval rigging. That substitute was offered by George Wright Binks, a foreman ropemaker in the Woolwich Dockyard. Binks conceived the idea of twisting together fine iron wire to form a rope, a specimen of which he submitted to the Admiralty. Too advanced, the idea was not approved. It would take Captain George Harris, an officer of the Royal Navy, to sell the innovation.

Captain Harris saw "the germ of a great and important manufacture" in Binks' wire rope specimen. Taking the ropemaker in tow, and with aid from Andrew Smith, a Scot mechanic who dabbled with wire rope, Harris established a small company at Great Grimsby in 1835. Wire rope for standing rigging of ships was their sole product, and to prove its worth they fitted out the ship *Marshall* as an experiment. The Admiralty bought the idea and proclaimed the merits of wire rope.

Rope made of wire was cheaper and lighter than hemp rope. One pound of iron is equivalent to one and three-quarters pounds of hemp, and the cost of the first was fourteen cents as against seventeen cents for the latter. Iron rigging did not have to be "set up" (tightened) after variations in weather, as hemp rope

Andrew Smith Hallidie was born in London, England, on March 16, 1836, the son of Andrew Smith of Fleming, Dumfrieshire, Scotland, and Julia Johnstone Smith of Lockerbie, Dumfrieshire. After coming to California, young Andrew Smith adopted the surname Hallidie in honor of his godfather and uncle, Sir Andrew Hallidie.

Like his father, Hallidie was an inventive genius. He established wire rope manufacture on the Pacific Coast, which greatly benefited Western mining, particular the Comstock. His cable railway system revolutionized public transportation worldwide, the royalties from which made him a wealthy man. He died of heart disease at age sixty-five, in San Francisco, on April 24, 1900.

had to be. Iron rope could be spliced and fitted in all kinds of weather, quicker, neater, and better than hemp. Unfortunately for Harris, Smith, and Binks wire rope was an invention that no one company could monopolize, and they soon faced enormous competition.

Within thirty years there were at least forty-five manufacturers in Britain, employing 7,000 men, turning out annually 17,000 tons of wire rope in a multitude of forms, from one to many strands, in gauges as fine as a hair to as thick as a finger. Wire was drawn from several metals: from cast, homogeneous, and Bessemer steel, from charcoal and common brands of iron, bright or annealed, and from copper and brass. Besides ships' standing and running rigging, wire was employed in telegraph cables, suspension bridges, for all sorts of engineering purposes, pneumatic telegraphs, traction ropes for tramways, steel plow ropes, coulippe ropes for transport of sugar cane, tent stays, tow ropes, endless driving bands, and dozens of other uses.[4]

While there were many Americans experimenting with wire rope between 1840 and 1870, three men are recognized for applying the invention to American industrial needs: John A. Roebling and his son William A., and Andrew S. Hallidie.

John A. Roebling, a New York inventor who held numerous patents for wire rope fabrication, made the first wire rope in the United States at Saxonburg, Pennsylvania, in 1840. It was a selvagee rope 500 feet long and three-quarters of an inch in diameter. That style of rope failed, and he next fabricated a 1,500 foot long rope, 1½ inches in diameter, that was durable and successful. For the next ten years the Roeblings engaged in suspension projects. In 1855 father and son constructed a railway bridge across Niagara Gorge. Thirty years later the building of the Brooklyn Bridge forever linked the Roebling name to architectural history. Although Roebling's counterpart in the West, Andrew S. Hallidie, also constructed suspension bridges, his fame would come from putting wire rope to use in mining and public transportation — and San Francisco would be his base of operation.

William A. Roebling and Andrew S. Hallidie had similar backgrounds. Both came from families engaged in wire rope manufacture. There was only a year's difference in their age, Hallidie being born in Dumfries, Scotland, on March 16, 1836, the son of Andrew Smith, who obtained on January 12, 1835, a patent for "a new standing rigging for ships and vessels and a new method of fitting and using it." This is the same Smith, who with Binks and Harris established a wire rope works at Great Grimsby. On March 26, 1836, Smith obtained a patent "for an improved mode of manufacturing bands, belts and straps to be employed in place of ropes and chains and for other useful purposes." Other patents followed in rapid succession: December 21, 1836, March 20, 1839, and later.[5]

These first ropes were made from a number of small wires set up on spools back of a stationary laying plate containing a series of holes arranged in

Another general view of Hallidie's endless wire rope-way, conveying ore from the top of a mountain in a rough mining region to the mill below. The wire rope passed over pulleys elevated upon posts, high enough so that the containers cleared all obstructions on the ground. As shown in the engraving, hand operated or self-dumping ore boxes, or for that matter ore sacks or cars, were suspended from the rope by devices which kept them relatively level. The boxes were loaded on the mine dump, and the grip pulley, an ingenious device, being revolved by means of the engine. The boxes passed down on one side, delivered their load, and passed up empty on the other side. The rope-way was run by the mill's engine; no extra power was needed, gravity of the descending load being sufficient to keep the rope-way in motion. Hallidie Endless Wire Rope-Ways were widely used in the intermountain West, particularly Utah, Idaho, Montana, and Colorado, and in the lumber regions of the Pacific Northwest. (*Mining and Scientific Press*, February 3, 1872)

circles around a center hole (i.e., six holes around one, then twelve around six and eighteen around twelve, etc.). Wires were pulled through these holes and formed into strands by a revolving machine which receded from the laying plate along a track on a long rope walk. These strands were afterwards laid up into ropes in a similar manner to hemp ropes. Modern wire rope manufacture was born about 1845 when Smith invented a machine for making wire rope strands of untwisted wires.

Young Andrew Smith acquired knowledge of wire rope making from his father, and honed his mechanical skills in his brother's machine shop. He was destined to put both to work in far-off California.

In the early 1850s Andrew Smith, Sr., became enarmoured with the prospects of gold mining on Fremont's Mariposa Estate, in California, shares of which were then being peddled in England. Andrew Smith apparently was impressed enough to look into the scheme, and on January 28, 1852, he and his son left Liverpool aboard the steamship *Pacific,* bound for New York. They sailed to Panama aboard the poorly equipped *Brother Jonathan.* In mid-March they crossed the Isthmus and embarked for San Francisco abroad the *Brutus,* arriving in San Francisco fifty-nine days later.

Andrew, Sr., and his son, inspected the Mariposa mines. Father, however, became disillusioned and returned to England, leaving his son to fend for himself in the gold camps of California. For the next three or four years, young Smith mined and did blacksmithing in Calaveras, El Dorado, Placer, and Nevada counties.

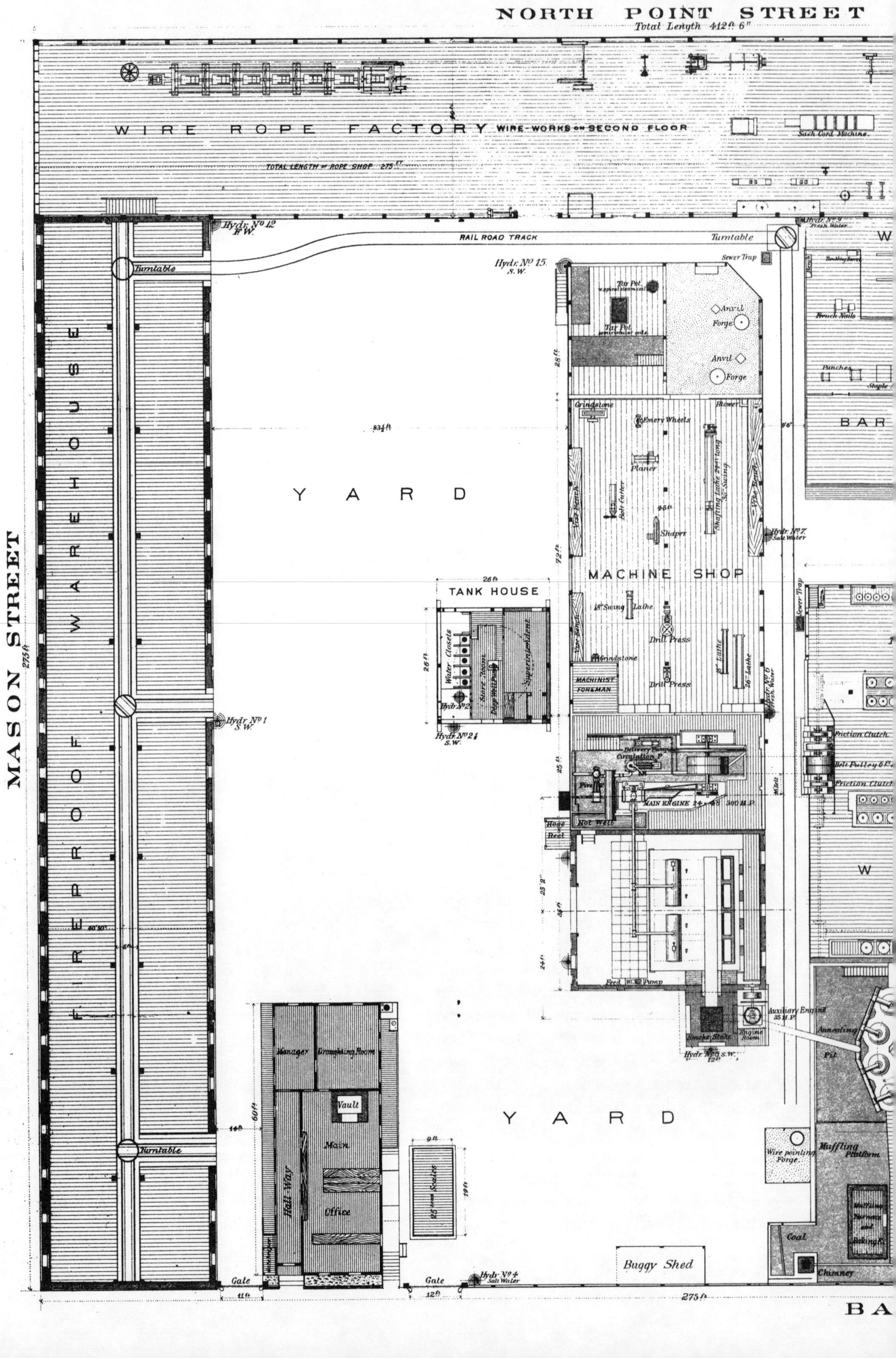

NORTH POINT STREET
Total Length 412 ft. 6"
WIRE ROPE FACTORY
WIRE-WORKS ON SECOND FLOOR
TOTAL LENGTH OF ROPE SHOP 275 ft.
Sash Cord Machine.
RAIL ROAD TRACK
Turntable
Hydr. No 12 F.W.
Hydr. No 15. S.W.
Sewer Trap
Tar Pot
Anvil
Forge
Grindstone
Emery Wheels
Planer
Shaper
Drill Press
Lathe
MACHINE SHOP
MACHINIST FOREMAN
TANK HOUSE
Water Closets
Store Room
Superintendent
Hydr. No 1 S.W.
Hydr. No 7. Salt Water
Hydr. No 24 S.W.
MAIN ENGINE
Hot Well
Hose Reel
Delivery Pump
Circulation P.
Fire P.
Feed Pump
Smoke Stack
Engine Room
Auxiliary Engine
Friction Clutch
Belt Pulley
Punches
Annealing Pit
Muffling Platform
Wire pointing Forge
Coal
Chimney
YARD
FIREPROOF WAREHOUSE
MASON STREET
275 ft.
Manager
Vault
Main Office
Hall Way
Scales
Buggy Shed
Gate
Hydr. No 4 Salt Water
BAR

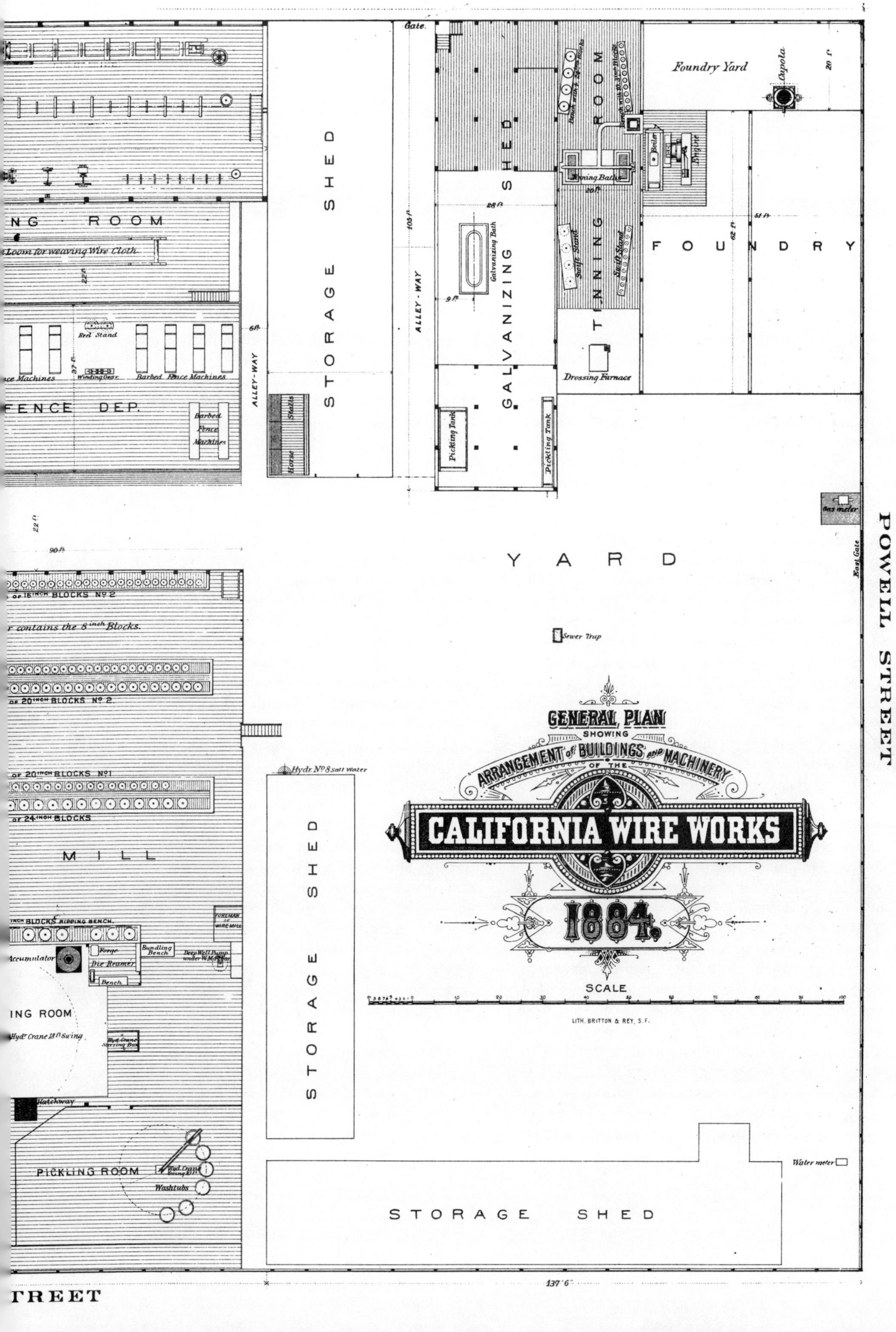
GENERAL PLAN
SHOWING
ARRANGEMENT of BUILDINGS AND MACHINERY
OF THE
CALIFORNIA WIRE WORKS
1884.
SCALE
LITH. BRITTON & REY, S.F.
POWELL STREET
TREET
YARD
STORAGE SHED
STORAGE SHED
STORAGE SHED
ALLEY-WAY
ALLEY-WAY
GALVANIZING SHED
TINNING ROOM
FOUNDRY
Foundry Yard
Cupola
Boiler
Engine
Tinning Baths
Galvanizing Bath
Pickling Tank
Pickling Tank
Dressing Furnace
Gate
East Gate
Gas meter
Water meter
Sewer Trap
Hydr. No 8 salt water
NG ROOM
Loom for weaving Wire Cloth
Reel Stand
Winding Gear
Barbed Fence Machines
FENCE DEP.
Barbed Fence Machines
Horse Stalls
OF 16 INCH BLOCKS No 2
contains the 8 inch Blocks.
OF 20 INCH BLOCKS No 2.
OF 20 INCH BLOCKS No 1
OF 24 INCH BLOCKS
MILL
INCH BLOCKS RIBBING BENCH.
FOREMAN OF WIRE MILL
Accumulator
Forge
Die Reamer
Bench
Bundling Bench
Deep Well Pump under W. M. Office
ING ROOM
Hydr. Crane 18 ft Swing
Hyd. Crane Steering Box
Hatchway
PICKLING ROOM
Washtubs
137' 6"

Fig. I.

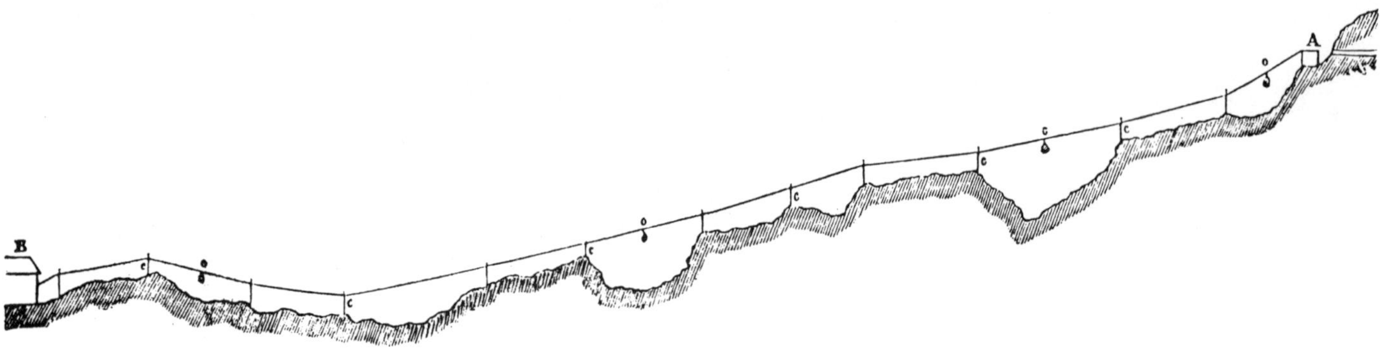

Above and at right: The components of Hallidie's Endless Wire Rope-Way. Fig. 1 (above) shows a section of rough mining region, over which it is desired to transport ore from the mine, A, to the mill, B, distant, say, one mile or more. At proper points, from 200 to 600 (usually) 250) feet apart, are erected posts, c, with guide-sheeves, on which the rope travels. For the sake of illustration, sacks, o, are shown in the cut, suspended by proper devices to the rope. Near A and B at each end of the rope-way, are placed horizontal grip-pulleys, devised by Hallidie, eight to twelve feet in diameter, around and in the groove of which runs an endless wire-rope of suficient length to extend from one pulley to the other and back, so that full sacks, or cars, can be run down on one side and back on the other.

Previous facing pages: The shops of the California Wire Works were a prominent feature of North Beach and could be observed from the decks of steamers entering the harbor. The site occupied an entire block, bounded by Bay, Powell, North Point, and Mason streets, more than two and a half acres of ground reclaimed from the bay, and filled to a height of twenty-two feet above the bed of the salt water.

The California Wire Works bought the property in September 1882, and in October commenced laying the foundations for their plant which included a wire mill, galvanizing works, wire rope factory, barbed wire factory, wire works, wire weaving shops, machine and fitting shops, foundry, carpenters' and pattern making shops, store rooms, and warehouses. In all, $680,000 were expended in real estate, equipment, construction, and in raw and manufactured material.

The company consisted of four divisions. The first, the machine shop, consisted of lathes, planers, shapers, drilling machines, bolt cutters, and associated blacksmith and pattern-making shops. All were powered by a 300 horsepower Corliss horizontal engine, manufactured by the Fulton Iron Works. The second department, the wire working and weaving shop, worked up wire into every conceivable shape, from massive gates and guards for prisoners and insane asylums, to delicate flower-baskets for a lady's conservatory. The wire rope factory manufactured long, massive flat wire ropes used for hoisting ore from deep mines. The main rope factory, where wire was spun into strands and the strands woven into rope, occupied a building 275 feet long by 40 feet wide, filled with ingenious and somewhat noisy machinery, all invented by A. S. Hallidie or his brother. Here great attention was given to the fabrication of cables for street railways. Barbed wire fence making was carried on in an adjacent building.

The various branches of the California Wire Works, as well as demand from outside companies, was supplied with wire from wire mills situated adjacent to the works. There, all wire was drawn cold, through steel dies having holes in them of the proper size and form needed for the purpose. In order to fit these dies, solid rods of Bessemer, Siemen-Martin, and Swedish iron were annealed in specially built furnaces. After annealing, the rods were subjected to acid baths to remove all oxide and scale. The rods were then covered with a gelatinous coating, and submitted to the wire drawers. This massive machinery consisted of eight different trains of revolving blocks, in all 236 blocks, through which all sizes of wire were made. The wire was then cleaned with acid, washed, and dried in a brick bake house. Wire that needed to be galvanized was sent to the company's mill and foundry on Powell Street. All lifting of wire and rods was done by hydraulic cranes.

The California Wire Works was one of the leading and profitable industries of the Pacific Coast, and furnished employment to 150 men, many of whom earned as high as $30 per week at piecework. (*Mining and Scientific Press*, August 2, 1884)

Fig. 2 shows construction of supporting posts, which have a cross-beam holding revolving grooved sheeves, a, b. The lower sheeve, b, supports the rope, the upper one, a, keeps it from jumping out of place. Fig. 6 shows the grip pulley employed at each end of the line. Fig. 7 is a section of the rim of the pulley, showing mode of construction. The rope, h; ii, are clips working on a fulcrum, jj. The rope pressing on the clips at the bottom, as it enters them, causes them to close over it, gripping it securely and preventing slippage.

Attached to the wire rope at equal distances apart, usually about every fifty feet, are peculiarly constructed carriers. Figs. 3 and 4 show these and the manner of attaching them to the wire; Fig. 3 giving a side view and Fig. 4 an end view. Here a and b are the sheeves, o the rope, and C, the carrier. This carrier is placed in a frame, d, which is hung on standards, e, in such a manner that the carriers will always maintain a horizontal position.

A carrier is shown in Fig. 8, found useful in certain cases, as it economized manual labor. A car, n, is mounted on wheels so that it could be run into the mine. It had a carrying frame, p, above it, the longitudinal beam of which was inclined to correspond with that of the standards, r. Both were toothed, the former on its lower side, the latter on its upper side. The standards, r, attached to the rope, came around, caught and carried off the car without use of manual labor.

Fig. II.

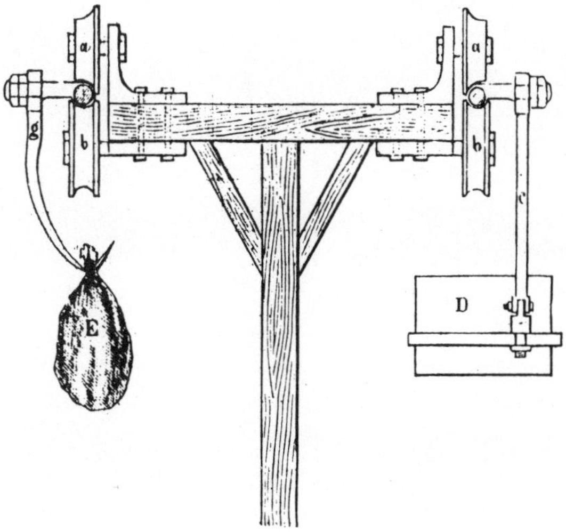

Fig. VI.

Fig. VII.

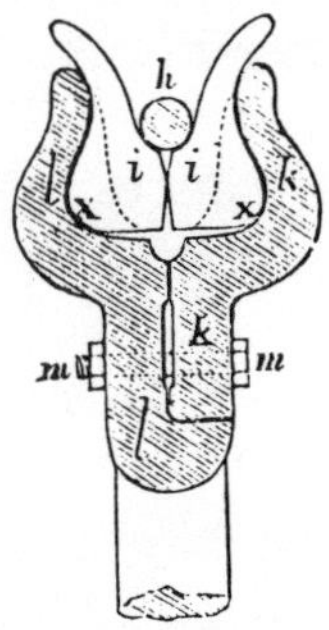

Fig. III.

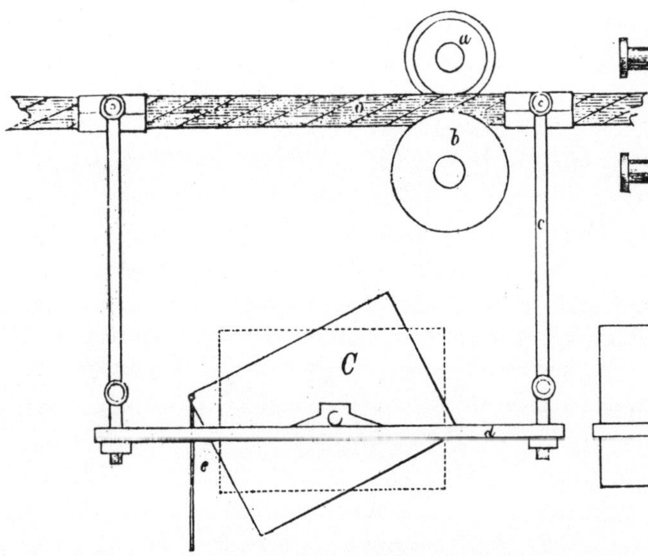

Fig. IV.

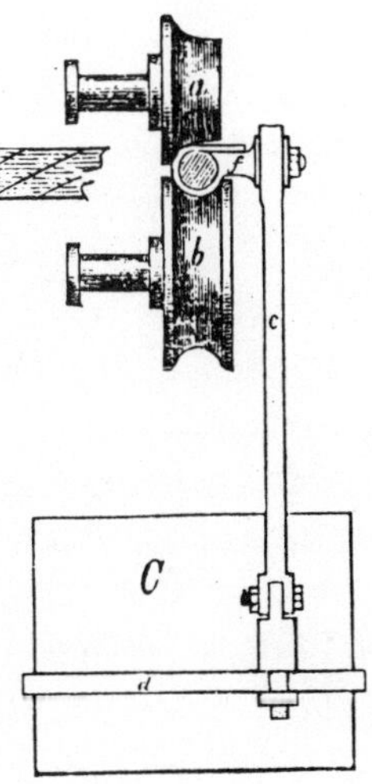

Fig. VIII.

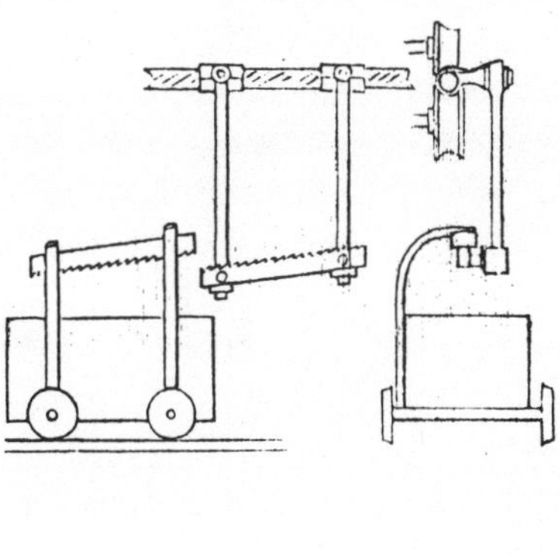

The standards, e, were attached to a bar, f, shown on a larger scale in Fig. 5 (below). The end of the bar was swaged out into a band which encircled the wire rope and was riveted to the bar, so as to hold the rope tightly. The bar, f, was about one half the size of the rope, and as the center of gravity of the load came vertically below the wire, the bar always stood out horizontal, thus allowing the load to be carried past the sheeves and pulleys without interference. On one end of the carrier was an apron, e, which enabled the load to be easily dumped.

Fig. V.

He also engaged in construction projects in the Mother Lode country. Drawing upon the skills he learned from his domineering father, Andrew, Jr., in 1855, when only nineteen years old, put a wire suspension bridge and an open flume, 220 feet long, across the middle fork of the American River, at a spot called Horse Shoe Bar.[6]

About this time Andrew, Jr., dropped the use of his last name. His reason for doing so has never been clarified. Resentment may have been the cause. Although Andrew never stated as much, walking in his father's footsteps was tough. Instead of Smith, Andrew adopted the surname of a famous uncle, Sir Andrew Hallidie, noted Scottish physician to King William IV and Queen Victoria. In so doing, Hallidie stepped out of his father's shadow to emerge as a master wire rope maker.

In 1856 Andrew Hallidie made the first wire rope ever produced on the West Coast for the Bay State Mine on the American Bar of the middle fork of the American River. The wire was bright iron, which Hallidie carbonized by treating it in a charcoal pit for several days. The rope was 2,200 feet long, two and a half inches in circumference, and was made to take the place of a hemp rope seven inches in circumference, which wore out every four and a half months. The rope which Hallidie made — with machinery fabricated by himself on the spot — was in three sections and spliced together.[7] It lasted two years. A year later, in 1857, Hallidie established a wire rope plant on Taylor Street on North Beach in San Francisco. Under the name of A. S. Hallidie & Company, he manufactured wire rope in sizes of No. 6 to No. 31, cords from one to sixteen inches in diameter, and two inch rope. Shortly after opening his company, Hallidie produced a rope measuring 2,275 feet long without a joint or splice. This naturally led to suspension projects.

Hallidie was building suspension bridges in the Far West a year before the Roeblings spanned Niagara Gorge. By 1869 A. S. Hallidie & Company had built the Folsom bridge, a 330 foot span; a 320 foot bridge at Nevada City; the Klamath River bridge, a 312 foot span; the Bear River bridge, 225 feet; the Frazer River bridge, 292 feet, as well as small spans in Oregon and Idaho territories.

The transition from scratching wealth from surface deposits to sinking deep shafts, forced Western mining companies to look for an alternative to hemp rope. Being in the right place at the right time, A. S. Hallidie & Company met that need with a variety of products: haulage or transmission rope, consisting of six strands of seven wires each; and high tensile strength haulage rope of six strands of nineteen wires each, which was flexible enough to wrap around drums and sheaves.

It was production of flat iron rope, however, that links A. S. Hallidie to the world of mining. In September 1863 he made the first flat iron rope for the Sierra Nevada Mine, on the Comstock. It was woven in two coils, each 700 feet long, and was four inches in breadth by half an inch in thickness. Warranted to raise ten tons, this rope was a great improvement over round iron rope than in use. It had higher tensile strength, wound better on drums, and was less liable to slip. A cage could thus be lowered more steadily. Flat rope was also less prone to wear from abrasion produced by sheeves and hoisting mechanisms. Cables of similar design, but made of braided steel wire, were later adopted by the Savage, Hale & Norcross, and a majority of other mines on the Comstock.[8] In 1869 the Ophir Mine placed an order with Hallidie for two flat wire ropes, five inches wide, three-eighths inch thick, and 1,500 feet long. These ropes proved their worth in deep hoisting, and for years A. S. Hallidie & Company was the sole source.[9]

With mining customers worldwide, a maritime business spanning the Pacific Rim, and a local trade in all sorts of wire mesh, and fencing wire, A. S. Hallidie & Company grew rapidly to become one of San Francisco's largest manufacturers. In 1870 A. S. Hallidie absorbed his only competition on the West Coast, Joshua Gray's Wire Rope Company, founded in 1866. On May 27, 1870, the Pacific Wire and Wire Rope Company was incorporated. Capital of the company consisted of $100,000 in 1,000 shares, with only $20,000 paid in as of March 1872. The trustees were James B. Stetson (president), Jonathan Kittredge, Robert Watt, David Farquharson, Charles F. Marwedel, J. M. Eckfeldt, Thomas Nelson, Edward Kruse, Joshua Gray, and A. Furhman. Andrew Hallidie acted as agent for the company, and Gray supervised manufacturing.[10]

The new firm was situated on Brannan Street, between Third and Fourth, the site of Joshua Gray's original plant. The two story main building, occupying a space 35 x 135 feet, housed three annealing furnaces, two of which had a capacity of 3,500 pounds each. With twenty employees and power supplied from a forty horsepower engine in an adjoining building, the company commenced operations in October 1871, with about twenty tons of wire being sold a month. Thirty-eight drawing blocks turned out all kinds of wire, from the largest size of three-eighths of an inch down to No. 36, which is .004 of an inch in thickness. Initially the firm concentrated on supply-

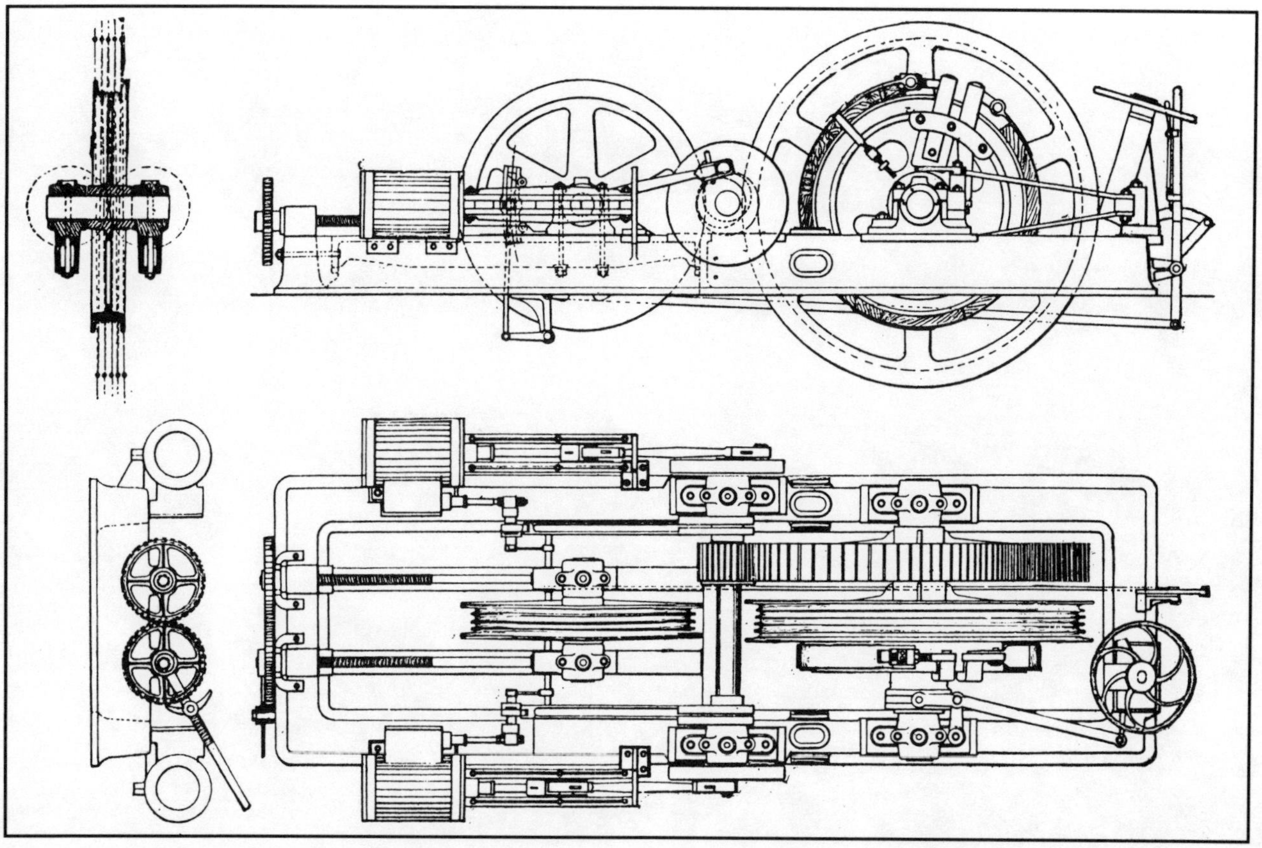

Above: An endless-rope hauling engine for mining purposes, consisting of two tandem placed drums. The rope is first passed over the larger and then over the smaller drum, as many times as may be found requisite to secure the necessary tractive force. The motion of the larger, or hauling drum, is controlled by a friction-clutch. The motive-power furnished by two engines, actuating a piston meshing with a spur-wheel keyed to the drum-shaft. The "idle" drum rests on sliding bearings, and slack rope may be taken up by draw-bolts, as shown. In the endless-rope system the amount of artificial tension introduced to prevent rope slippage upon the drums had to be considered. The tension depends, of course, upon the number of times the rope is passed over the hauling drum. (*Mining and Scientific Press*, August 3, 1889)

ing broom, spring, wire rope, and wire cloth, as well as baling and fencing wire.[11] What really propelled the company to the forefront of West Coast business was Hallidie's Endless Wire-Rope-Way.

Although Hallidie patented this invention in 1869 through the patent service of Dewey and Company, publishers of the *Mining and Scientific Press*, endless rope-ways had been around since the Middle Ages; and the deep mines of Britain began using them in the mid-1850s. Nevertheless, being the only supplier of wire rope in Western America, Hallidie's invention revolutionized not only ore haulage, but public transportation as well. Since the focus of this book is mining, we'll look at what Hallidie did for that industry before mentioning his achievements in public transportation.

Hallidie's Endless Wire Rope-Way consisted of iron or steel wire ropes, supported on sheeves placed on posts, actuated by the gravity of descending loads, or by an engine attached to a grip pulley. Every fifty feet along the rope were carriers, placed in such a way that they always maintained a horizontal position. By means of these carriers, which ran at 200 feet per minute, ore, timber, goods, just about anything, could be transported over rough or inaccessible country. Delivering twelve tons per hour, the system was a boon to mining companies.[12]

Gravity propelled rope-ways had their drawbacks. Because they were endless, wrapping around large wheels, often six feet or more in diameter, their inertia could provide more power than was actually needed to keep them in motion. Without some sort of brake or brakes the tramways could run away, destroying the entire system. The key to the Hallidie Endless Rope-Way were brake drums of a diameter equal to that of the bull wheel, incorporated at each

Hallidie was not the only manufacturer of wire ropeways. Parke and Lacy sold the Huson tramways throughout the West. They installed the above Huson tramway at the Pay-Rock Mine, Silver Plume, Colorado. This tramway, built on a twenty-one degree angle, carried ore from the mine, and transported coal and tools back to the workings, in every third ascending bucket. (*Mining and Scientific Press*, January 1, 1887)

terminal. These brakes were so powerful the cables often slipped on the wheel surfaces. Hallidie remedied that by installing a series of clamping pairs in the pulley wheel at intervals of six inches. Hinged on a center pin, these clamps pressed the cable against the bottom of the clamping fingers. At the same time, the top halves of the clamps folded around and tightly gripped the cable. With release of pressure as the cable left the wheel, the clamps' tops popped open, releasing the cable.

About the time Hallidie was employing this gripping mechanism on tramways carrying all sorts of cargos, he was applying it to urban transportation. According to Walt Wheelock, the inventive Scot conceived the idea of a cable car one foggy night in 1869 while trudging up the steep hills of San Francisco. He came upon a heavily laden horsecar, the public transportation of that day, struggling to climb the steep grade of the street. The driver cursed and thrashed the animals to force them up the slippery cobblestone surface. Suddenly one of the horses fell to its knees. The driver frantically grabbed the brake control but it failed to hold, and the car rolled to the foot of the hill, cruelly dragging the hapless beasts behind it. From then on, so the story goes, Hallidie became a man with a mission dedicated to saving the thousands of horses destroyed each year by trams laboring up the hills of the world. Thus, San Francisco's cable car system was conceived. A good story, or a good press release to promote his wire rope business? Probably the latter, for endless wire rope ways had been applied to mining before public transportation.[13]

Until supplanted in the 1890s by the German engineered Bleichert & Co. tramways, Hallidie's Endless Rope-Ways transported ore from mine to mills and loading docks in mountainous areas. In many cases their use made mining profitable where traditional freighting was impossible. Hallidie Rope-Ways were installed in Utah at Little Cottonwood and Bingham Canyons, at Mineral King in California, and at numerous sites in Nevada, Washington, Oregon, and Idaho. A Hallidie Rope-Way was proposed in 1898 as a method of easing the passage of treacherous Chilcoot Pass in the Klondike. It was a Bleichert tramway that ultimately solved that problem, however.

At right: Upper terminal of the Huson system, showing manner of loading the buckets. As the bail of the bucket intersects with and is held by the sprocket or rag-wheel; the pendant of the bucket coming in contact with a stop projecting outward from the loading car, prevents the bucket from passing around the end, without moving the loading car with it, and as the loading car moves upon a circular track, the cam at the top raises the gate of the loading car by means of a grooved roller connected to the gate, and allows the load to drop directly into the bucket.

Illustration at left represents the lower terminal of the Huson system, and the mode of dumping the bucket. This lower terminal is pitched or inclined to conform with the mountain, and by means of the side car or cage the pendant is changed from a hanging or vertical position to that of right angles with the rope. When the pendant comes in contact with the cage, the lower end is held back by a projection on a tripping bar connected to or with the car, until the clip at upper end of pendant, which is fast to the rope, strikes or comes against a similar stop at the upper end of the tripping bar, then the cage is moved downward with the clip upon the rope until the bail of the bucket is fairly into the sprocket wheel; then the tripping bar and stops are released and the cage is returned with a weight ready for the next bucket. A series of cams actuates the bucket and dumps the ore into bins.

Rope-ways and flat wire rope were not the only commodities Hallidie furnished the mining industry. In 1883 his company introduced a new type of battery screen. Previous, screens were fabricated from a single piece of punched metal, or piece of wire cloth of the desired size of mesh. The Pacific Wire and Wire Rope Company introduced a screen with angles or corrugations, to give more screen surface to the given size of the battery opening. The improved screens were constructed of woven wire, fixed in frames shaped to fit the openings in the mortars.[14]

Last, but certainly not least, a few words about the man. As his obituary states, "To Mr. Hallidie, more probably, than to any other one man, does San Francisco owe its urban development. Prior to 1873 the trend of travel was suburban: he made possible rapid transit up the steep hills to the west of the then thickly settled parts of the city, the cable railway developing and building up that part of the peninsula between the bay and the ocean." His contributions to city and state did not stop there. In 1868 he was elected president of San Francisco's Mechanics' Institute, an office he held for ten years. He was also a prominent member and officer in nearly every local society for commercial or scientific advancement. For thirty years Hallidie served as a regent of the University of California, and he was a trustee of the California School of Mechanical Arts. Personally Andrew S. Hallidie was a quiet and unassuming man, who readily assisted anyone seeking his help. He took an active interest in everything pertaining to the industrial and public welfare of his adopted city. When he died on April 24, 1900, at age sixty-five, San Francisco marked the passing of one of her foremost citizens.[15]

Below: A novel scene on the street of Telluride, Colorado. Twelve thousand feet of wire rope, weighing 15,000 pounds, packed on sixty-five mules. The rope was destined for a tramway between the Nellie Mine and the Telluride Power Transmission Company's mill on Bear Creek. (*Mining and Scientific Press*, August 7, 1897)

Chapter Sources

Chapter One

1. Thomas A. Rickard, *A History of American Mining*. New York and London: McGraw-Hill Book Co., pp.20–23.

2. Herbert H. Bancroft, *History of California*. San Francisco: The History Co., Vol. XXIV, pp. 695-96.

3. Theressa Gay, *James W. Marshall. The Discoverer of California Gold*. Georgetown, California: Talisman Press, 1967, p. 225.

4. W. R. Dodge, "Gold Mining and Milling in the Southern States," *Mining and Scientific Press*, January 9, 1915. Hereafter cited as MSP.

5. L. White, *Medieval Technology and Social Change*. Oxford: Oxford University Press, 1962, p. 81.

6. N. P. Boss, "The Pachuca Stamp Battery and Its Predecessors," MSP, August 9, 1902.

7. White, *Medieval Technology*, pp. 83-84; Sandy Gerrard, "The Medieval and Early Modern Cornish Stamping Mill," *Industrial Archaeological Review*, Vol. XII, No. 1, Autumn 1989, pp. 9–10.

8. Herbert Hoover, (trans. and edit.) *De Re Metallica*. New York: Dover Publications, Inc., 1950, pp. 279–286.

9. *Ibid.*, pp. 282–83.

10. It is definitely known that in 1561 Queen Elizabeth imported "some Germans experienced in mines," and granted them mineral rights within the counties of York, Lancaster, Cumberland, Westmoreland, Cornwall, Devon, Gloucester, Worcester, and Wales. They utilized the noisy gravity stamp mills to crush a quartzose gangue to extract cassiterite, the oxide of tin. *Ibid.*, p. 283.

11. MSP, December 6, 1902; MSP, June 12, 1915.

12. Courteney De Kalb, "The California Stamp-Mill," MSP, May 21, 1910.

13. MSP, May 21, 1892.

14. MSP, September 8, 1860.

15. H. H. Bancroft. *History of California*. Vol. XXIV, p. 94.

16. A. D. Hodges, Jr., "Amalgamation at the Comstock Lode," MSP, February 14, 21, 28, 1891.

17. MSP, April 30, 1898.

18. Eliot Lord. *Comstock Mining and Miners*. San Diego: Howell-North Books, 1980, pp. 113-114.

19. MSP, April 18, 1898.

20. MSP, August 26, 1893.

21. Dan De Quille. *The Big Bonanza*. New York: Alfred A. Knopf, 1967, pp. 89-90; Lord, *Comstock Mining and Miners*, p. 221.

22. Lord, *ibid.*, p. 224.

23. MSP, December 22, 1888; see also Frank C. Roberts, "Wire Rope Haulage and Its Application to Mining," *Transactions of the American Institute of Mining Engineers*, Vol. XVL, 1887–88, pp. 213–262.

24. MSP, September 29, 1888.

25. MSP, May 8, 1875.

26. Dan De Quille, *Big Bonanza*, p. 122.

27. Walter D. O'Brien, "Recent Work on the Comstock," MSP, June 13, 1908.

28. P. B. McDonald, "Driving the Hoosac Tunnel and the Lessons it Taught," MSP, October 10, 1914.

29. George J. Specht, "On Rock Drills," MSP, April 26, 1884.

30. MSP, June 22, 1872.

31. MSP, February 15, 1873.

32. MSP, May 8, 1875.

33. Joseph T. Goodman, "Comstock Beginnings," MSP, July, 3, 1909.

34. Walter D. O'Brien, "Recent Work on the Comstock," MSP, June 13, 1908.

35. John Haskell Kemble. *San Francisco Bay. A Pictorial History*. Cambridge, Maryland: Cornell Maritime Press, 1957, pp. 7-12.

36. MSP, September 10, 1864

37. For details relative to the founding of the Mechanics' Institute Library see, Edgar M. Kahn, "Mechanics' Institute of San Francisco," in *Seven Pioneer San Francisco Libraries*, published by The Roxburghe Club of San Francisco, 1968, pp. 19–24.

38. Charles G. Yale, "Twenty-three Years a Mining Editor," MSP, May 21, 1910.

39. *ibid.*

40. Dexter, H. F., "Looking Backwards," MSP, May 21, 1910.
41. *Ibid.*
42. MSP, December 28, 1872; February 8, 1873.
43. MSP, December 15, 1866; H. F. Dexter, MSP, May 21, 1910.
44. L. A. Greene, "Fifty Years of Advertising," MSP, May 21, 1910.
45. H. F. Dexter, MSP, May 21, 1910.
46. MSP, April 6, 1878; September 13, 1879.
47. MSP, July 9, 1910.
48. MSP, March 23, 1895.
49. MSP, June, 10, 1893.
50. MSP, December 15, 1894.

Chapter Two

1. *Dictionary of American Biography.* 10 vols. Edited by Allen Johnson and Dumas Malone. New York: Charles Scribner's Sons, 1975, Vol. III, pp. 362–363.
2. *Ibid.*, Vol. VIII, pp. 491–92. For additional details on the life of Peter Donahue consult H. H. Brancroft. *Chronicles of the Builders of the Commonwealth.* San Francisco: The History Co., 1891, 7 vols. Also O. T. Shuck. *Sketches of Leading and Representative Men of San Francisco.* San Francisco: Beacon & Co., 1875.
3. MSP, July 21, 1888.
4. MSP, May 5, 1866.
5. MSP, July 21, 1888.
6. MSP, August 9, 1902.
7. MSP, May 5, 1866.
8. MSP, July 1, 1865; November 17; December 8, 1866.
9. MSP, August 25, 1866.
10. MSP, October 1, 1870; April 26, 1873.
11. MSP, April 13, 1867.
12. MSP, October 1, 1870.
13. MSP, June 20, 1874 and July 3, 1875.
14. MSP, June 6, 1866.
15. MSP, December4, 1886.
16. MSP, August 14, 1875.
17. MSP, February 26, 1876.
18. MSP, April 6, 1878.
19. MSP, February 14, 1880.
20. MSP, October 27, 1883; Ruth Teiser, "The Charleston: an Industrial Milestone," *California Historical Society Quarterly,"* Vol. XXV, No. 1, March 1946, pp. 39–53.
21. MSP, July 10, 1886.
22. MSP, December 4, 1886. The Union Iron Works built the *Charleston* for $1,132,000.
23. MSP, October 15, 1887.
24. MSP, December 15, 1888; May 9, 1891.
25. MSP, June 12, September 18, 1886.
26. MSP, October 28, 1893.
27. MSP, November 14, 1896.
28. MSP, January 28, 1899.
29. MSP, September 1860. According to Sam Ward, E. B. Goddard was a "spare and pale New Englander" who had a "tolerable knowledge of machinery, running gear and the steam engine." In 1851 Goddard had an interest in the Washington Mining Company near Quartzburg, California. Carvel Collins (ed.). *Sam Ward in the California Gold Rush.* Stanford: Stanford University Press, 1949, pp. 94–95.
30. MSP, April 30, 1864.
31. MSP, May 21, 1864.
32. MSP, September 12, 1860; September 9, 1865.
33. MSP, September 10, 1864; May 27, 1865.
34. MSP, June 2, 1877.
35. MSP, March 30, 1867.
36. MSP, February 26, 1870.
37. MSP, August 21, 1869.
38. MSP, May 18, 1867.
39. MSP, April 27, 1872.
40. MSP, September 19, 1874.
41. MSP November 23, 1874.
42. MSP October 8, 1870.
43. MSP, March 15, 1873.
44. MSP, June 20, 1874.
45. MSP, May 8, 1875.
46. MSP, June 26, 1875.
47. MSP, June 17, 1876.
48. MSP, July 20, 1878.
49. MSP, August 25, 1877.
50. MSP, April 6, 1878.
51. MSP, Sept. 10, 1898.
52. Lynn R. Bailey, *Bisbee: Queen of the Copper Towns.* Tucson: Westernlore Press, 1983, pp. 14–16, 20–22.
53. MSP, April 30, 1881.
54. MSP, January 14, 1882.
55. MSP, October 21, 1882.
56. MSP, March 4, 1882.
57. MSP, December 2, 1882.
58. MSP, September 25, 1880.
59. MSP, April 30, 1881.
60. MSP, June 26, 1880.
61. MSP, March 8, 1884.
62. MSP, March 8, 1884.
63. MSP, April 3, 1886.
64. MSP, May 15, 1886.
65. MSP, October 8, 1887.
66. MSP, March 26, 1887.
67. MSP, May 21, July 2, and December 31, 1887.
68. MSP, January 29, 1887.
69. MSP, May 31, 1890.
70. MSP, February 13, 1892.
71. MSP, February 13, 1892.
72. MSP, October 28, 1893.
73. MSP, March 10, 1894.
74. MSP, July 6, 1895.
75. MSP, March 21, 1896.
76. MSP, September 22, 1900.
77. MSP, January 12, 1901.

Chapter Three

1. MSP, March 18, 1865.
2. MSP, September 9, 1865.
3. MSP, June 9, 1860.
4. MSP, April 9, May 28, 1864.
5. MSP, September 10, 1864; January 21, June 24, 1865.
6. MSP, May 27, 1865.
7. MSP, September 10, 1864.
8. MSP, March 2, 1867.
9. MSP, December 7, 1867.
10. MSP, April 14, 1866.
11. MSP, August 14, 1869.
12. MSP, July 9, 1870.
13. MSP, August 3, September 21, 1872.
14. MSP, May 15, 1875.
15. MSP, September 8, 1860.
16. MSP, March 25, 1862.
17. MSP, September 16, 1865.
18. MSP, July 16, 1870.
19. MSP, June 3, 1871.
20. MSP, May 12, 1877.
21. MSP, August 25, 1877.
22. MSP, April 6, 1878.
23. MSP, September 10, 1881.

24. MSP, April 9, 1887.
25. MSP, August 5, July 15, 1882.
26. MSP, January 13, 1883.
27. MSP, November 23, 1872; February 15, 1873.
28. MSP, June 25, 1864.
29. MSP, August 3, 1867.
30. MSP, July 16, 1870.
31. MSP, February 15, 1873.
32. Colorado Steam Navigation Company was organized in 1869, MSP, 27, 1872. For additional details on Colorado River steamboating see Arthur Woodward, *Feud on the Colorado*. Los Angeles: Westernlore Press, 1955.
33. MSP, November 23, 1872.
34. MSP, October 20, 1877.
35. MSP, December 7, 1878.
36. MSP, September 13, 1879; February 5, 1881.
37. MSP, March 8, 1884.
38. MSP, July 10, 1886.
39. MSP, March 26, 1887.
40. MSP, April 17, May 15, 1886.
41. MSP, July 2, 1887; September 8, 1888.
42. MSP, June 22, 1889 and May 17, 1890.
43. MSP, December 10, 1887.
44. MSP, June 4, 1892.
45. MSP, October 8, 1892.
46. MSP, February 4; March 18, 1893.
47. MSP, November 18, 1893.
48. MSP, December 22, 1894.
49. MSP, May, 18, 1895.
50. MSP, October 16, 1897.
51. MSP, February 1, 1861.
52. MSP, April 27, 1872
53. MSP, October 21, 1871.
54. MSP, October 21, 1871.
55. MSP, November 16, 1872.
56. MSP, April 26, 1873.
57. MSP, June 20, 1874.
58. MSP, February 26, 1876.
59. MSP, July 15, 1876.
60. MSP, February 24, 1877.
61. MSP, August 5, 1882.
62. MSP, September 13, December 1, 1883.
63. MSP, September 13, 1879.
64. MSP, April 20, 1889.
65. MSP, March 5, 1892.
66. MSP, June 11, 1887; May 16, 1891.
67. MSP, April 20, 1889.
68. MSP, September 3, 1892.
69. MSP, March 4, 1893.
70. MSP, May 27, 1893.
71. MSP, October 12, 1895.
72. MSP, September 3, 1892.
73. MSP, April 26, June 7, 1890; September 22, 1900; April 23, 1892; April 6, 1901.
74. John H. Kemble, *San Francisco Bay. A Pictorial Maritime History*, p. 72.

Chapter Four

1. MSP, April 28, May 12, 1866.
2. MSP, November 16, 1878.
3. MSP, June, 2, 1894.
4. MSP, May 19, 1866.
5. MSP, December 14, 1867.
6. MSP, February 5, 1870.
7. MSP, August 14, 1869.
8. MSP, September 23, 1871.
9. MSP, May 12, 1877.
10. MSP, March 20, 1880.
11. MSP, February 11, 1882.
12. MSP, December 6, 1884.
13. MSP, November 18, 1899.
14. MSP, November 18, 1899; May 25, 1901.
15. MSP, February 5, 1876.
16. MSP, July 7, 1877.
17. MSP, October 12 1878.
18. MSP, March 12, 1881.
19. MSP, January 1, 1887.
20. MSP, December 31, 1887.
21. MSP, October 29, 1887.
22. MSP, September 24, 1898.
23. MSP, January 14, 1888.
24. MSP, February 4, 1888.
25. MSP, February 16, 1889.
26. MSP, April 20, 1889.
27. MSP, April 26, 1890.
28. MSP, September 28, 1889.
29. MSP, May 26, 1888.
30. MSP, December 31, 1892.
31. MSP, June 30, 1894.
32. MSP, June 9, 1894.
33. MSP, December 22, 1894.
34. MSP, June 29, 1895.
35. MSP, July 29, 1899.
36. MSP, January 19, 1907.
37. MSP, November 8, 1879.
38. MSP, November 13, 1886.
39. MSP, September 15, 1888; March 2, 1889.
40. MSP, May 28, 1864; September 9, 1865; April 9, 1864.
41. MSP, February 27, 1864.
42. MSP, July 31, 1869.
43. MSP, December 10, 1870.
44. MSP, November 23, 1872.
45. MSP, September 30, 1893.
46. MSP, May 17, 1893.
47. MSP, July 16, 1892; January 14, 1893.
48. MSP, Aug. 5, 1893; December 18, 1897.
49. MSP, March 12, 1887.
50. T. Kirke Rose, *The Metallurgy of Gold*, London: Charles Griffin and Company, 1894, pp. 110–11.
51. MSP, March 12, 1887.
52. MSP, March 19, 1881.
53. MSP, August 18, 1883; April 9, 1887.
54. MSP, May, 26, 1888
55. MSP, October 24, 1891.
56. MSP, May 4, 1889.
57. MSP, June 4, 1898.
58. MSP, September 23, 1893.

Chapter Five

1. MSP, May 21, 1910.
2. A. J. Wallis-Tayler, *Aerial or Wire Rope-Ways. Their Construction and Management*. London: Crosby Lockwood and Son, 1911, p. 1.
3. *Mines and Minerals*, Vol. 24, April 1904, p. 442.
4. MSP, August, 2, 1873.
5. MSP, August 16, 1884.
6. Edgar M. Kahn, "Andrew Smith Hallidie," *California Historical Society Quarterly*, Vol. XIX, No. 1, March 1940, pp. 145–56; see also Kahn's article "Andrew S. Hallidie as Writer and Speaker," *California Historical Society Quarterly*, Vol. XXV, No. 1, March 1946, pp. 1–16. MSP, August 16, 1884.
7. Eliot Lord, *Comstock Mining*, p. 224.
8. MSP, October 23, 1869.

9. MSP, October 23, 1869.
10. MSP, June 11, 1870.
11. MSP, March 9, 1872.
12. MSP, February 18, 1871; September 5, 1874.
13. Walt Wheelock (ed.), *The Archaeology of the Cable Car.* Pasadena: Socio-Technical Books, 1970, pp. 4-9.
14. MSP, September 15, 1883.
15. MSP, April 28, 1900.

Collateral Sources

Appleton's Cyclopaedia of Applied Mechanics: A Dictionary of Mechanical Engineering and the Mechanical Arts. New York: D. Appleton and Company, 1880. 3 vols.

Appleton's Cyclopaedia of American Biography. New York: D. Appleton & Co., 1887–1931.

Behr, Hans. *Mine Drainage, Pumps, Etc.* Sacramento: California State Mining Bureau, 1896.

Chalfant, W. A. *The Story of Inyo.* Bishop: by author, 1933.

Collins, Henry F. "Cornish Pumping Engines," *Mining and Scientific Press,* February 20, 27, 1909.

DeGroot, Henry. "The Comstock Papers," *Mining and Scientific Press.* A series of 20 articles published between May 6, 1876 and March 24, 1877.

———. "Recollections of California Mining Life," *Mining and Scientific Press,* 5 installments, November 3 thru December 15, 1883.

Drinker, Henry S. *Tunneling, Explosive Compounds, and Rock Drills.* New York: John Wiley and Sons, 1888.

Gauvain, W. Percy, "Compound Cornish Pumping Engines," *Mining and Scientific Press,* July 10, 1909.

Gerrard, Sandy, "The Medieval and Early Modern Cornish Stamping Mill," *Industrial Archaeology Review,* Vol. XII, No. 1, Autumn 1989, pp. 9–19.

Hackett, Fred H. (ed.). *The Industries of San Francisco, Her Rank, Resources, Advantages, Trade, Commerce, Manufacturing.* San Francisco: Payot, Upham & Co., Publishers, 1884.

Hall, Leon M., "Comstock Drainage Problems," *Mining and Scientific Press,* July 3, 1909.

Hammond, John Hays. *The Autobiography of John Hays Hammond.* Murray Hill: Farrar & Rinehart, 1935. 2 vols.

Hodges, A. D., "Amalgamation at the Comstock Lode," *Mining and Scientific Press,* a series three articles appearing between February 14 and 28, 1891.

Knight, Edward H. *Knight's American Mechanical Dictionary. A Description of Tools, Instruments, Machines, Processes, and Equipment; History of Invention, General Technological Vocabulary; and Digest of Mechanical Applications in Science and Their Arts.* New York: Hurd and Houghton, 1876. 3 vols.

Liggett, J. H. (ed.). *The Industries of San Francisco, California. A Review of the Manufacturing, Mercantile and Business Interests of the Bay City, Together with a Historical Sketch of Her Rise and Progress.* San Francisco: Cosmopolitan Publishing Co., 1889.

Morrell, W. P. *The Gold Rushes.* London: Adam and Charles Black, 1968.

Palmer, Marilyn and Peter Neaverson. "Nineteenth Century Tin and Lead Dressing: A Comparative Study of the Field Evidence," *Industrial Archaeology Review,* Vol. XII, No. 1, Autumn 1989, pp. 20–39.

Peters, Edward Dyer, Jr. *Modern Copper Smelting.* New York: The Engineering and Mining Journal, 1903.

Preston, Edward B. *California Gold Mill Practices.* Sacramento: California State Mining Bureau, 1895.

Read, Thomas T., "Impressions of the Comstock," *Mining and Scientific Press,* August 24, 1912.

Richards, Robert H. *Ore Dressing.* New York and London: The Engineering and Mining Journal, 1906. 2 vols.

Ross, G. M. "The Great Comstock Lode," *Mining and Scientific Press,* October 12, 1907.

Shumate, Albert. *A San Francisco Scandal. The California of George Gordon.* San Francisco: California Historical Society, 1994.

Symmes, William. "Decline and Revival of Comstock Mining," *Mining and Scientific Press,* October 10, 24, 1908.

Wellington, C. Wolpe (ed.). *Men of California.* San Francisco: The Pacific Arts Co., 1901.

West, H. E., Cornish Tin Mining, *Mining and Scientific Press,* February 9, 1907.

Williams, David. *The Georgia Gold Rush: Twenty-Niners, Cherokees, and Gold Fever.* Columbia: University of South Carolina Press, 1993.

Young, Otis E., Jr. *Western Mining. An Informal Account of Precious-Metals Prospecting, Placering, Lode Mining, and Milling on the American Frontier from Spanish Times to 1893.* Norman: University of Oklahoma Press, 1970.

———. *Black Powder and Hand Steel. Miners and Machines on the Old Western Frontier.* Norman: University of Oklahoma Press, 1976.

Index